MODULAR SYSTEM

JOSEPH WEST

HUMAN BIOLOGY

PREFACE

Biology is a rapidly developing branch of science. The major advances that are made, continuously affect our life on earth. Some of these important advances are included here.

The results of a recent survey on the attitudes to existing literature available to high school students showed that many were unhappy with the material used in teaching and learning. Those questioned identified a lack of the following; accompanying supplementary material to main text books, current information on new developments, clear figures and diagrams and insufficient attention to design and planning of experiments.

This book aims to improve the level of understanding of modern biology by inclusion of the following; main texts, figures and illustrations, extensive questions, articles and experiments.

Each topic is well illustrated with figures and graphs to ease understanding. Supplementary material in the form of posters, transparencies and cassettes will shortly be available.

Profiles on common diseases are included in each chapter to inform, generate further interest and encourage students to explore the subject further. The 'Read me' articles supply up-to-date information on important issues related to each unit but outside the requirements of the current curriculum.

It is the intention and hope of the authors that the contents of this book will help to bridge the current gap in the field of biology at this level.

THE AUTHORS

CONTENTS

**Chapter 1: Basic Concepts of
        Anatomy and Physiology**
Body Organization of Organisms **6**
    1. Atoms - Molecules - Compounds 6
    2. Cells 6
    3. Tissues 7
    4. Organs And Systems 8
Human Physiology **8**
    1. Body Cavities 8
    2. Fluids Of The Body 8
    3. Membranes of the Body 9
    4. Homeostasis 9

Chapter 2: Nervous System
The Nervous System **16**
    1. Organization of the Nervous System . .17
    2. The Human Nervous System 22

Chapter 3: Senses
Sensory Reception **38**
Receptors And Sensation **38**
The Human Eye **39**
    1. The structure of the eye sphere 39
    2. External Structures of the Eye 41
    3. Visual Process 42
    4. Eye Defects 43
The Ears **45**
    1. The Structure of the Ears 45
    2. Hearing .49
Skin **52**
    1. Epidermis .53
    2. Dermis 54
    3. Accesory Structures of the Skin 54
    4. Touch Receptors 56
Smell **57**
    1. Mechanism of Smelling 58
Taste **60**
    1. The function of the tongue 61

Chapter 4: Endocrine System
Endocrine System **64**
    1. Hormones According to
       Their Chemical Structure 64
    2. Target Organs of Hormones 65
    3. The Regulation of Hormone Secretion 66
The Human Endocrine System **67**
    1. Endocrine Glands
       in the Human Body 67

Chapter 5: Locomotion Systems
Locomotion Systems **84**

Human Skeletal System **85**
    1. Bone Formation and
       Its Regulation 85
    2. Types of Bones 87
    3. Bone Growth 88
    4. Parts of the Human Skeleton 90
    5. Joints 95
Muscles **99**
    1. Muscle Tissue 99
    2. The Muscular System
       of Vertebrates 102

Chapter 6: Circulatory System
Circulatory System **112**
The Human Circulatory System **112**
    1. Heart 112
    2. Blood Vessels 119
    3. Blood 125
Immunology **135**
    1. Organs Of The Immune System 135
    2. Acquisition of Immunity 137
    3. Types of Immunity 137
    4. Allergy .144
    5. Immunologic Tolerance 144
    6. Vaccines 144
    7. Serum 145

Chapter 7: Respiratory System
Respiratory Systems **150**
    1. The Human Respiratory System 150

Chapter 8: Digestive System
Nutrients and The Digestive System **164**
    1. Digestive Systems 164
The Human Digestive System **166**
    1. Organs of the Digestive System 167
    2. Digestive Secretions 173
    3. The Digestion of Food 179
    4. Absorption 181
Nutrition **186**
    1. Carbohydrates 186
    2. Lipids 187
    3. Proteins .187
    4. Vitamins 188
    5. Minerals 189
    6. Water 189

Chapter 9: Excretory System
Excretory System **196**
    1. Excretory Substances 196
    2. The Human Excretory System 197
    Glossary **209**

BIOLOGY
HUMAN

Basic Concepts of Anatomy and Physiology

chapter 1

FIGURE: An atom

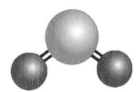

FIGURE: A molecule

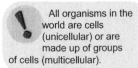

All organisms in the world are cells (unicellular) or are made up of groups of cells (multicellular).

BODY ORGANIZATION OF ORGANISMS

Organisms develop from a zygote formed by the fertilization of an egg by a sperm. A living organism can carry out all its biological functions, survive independently, grow and develop during its life span and reproduce to form offspring.

An organism includes six groups of components (in order of size): atoms, molecules, cells, tissues, organs and systems. These components are closely related to each other for the survival of the organism.

Anatomy deals with the structures that constitute the organism, while physiology deals with the function of these structures.

1. Atoms - Molecules - Compounds

Atoms are the simplest structural elements of living things, and more than 100 elements have been identified in the world. Each element has a unique atomic structure. The most frequent elements found in living things are carbon, hydrogen, oxygen, nitrogen, phosphorus and sulphur. Two atoms of the same element generally associate to form a molecule. For instance, two hydrogen atoms associate to form a hydrogen molecule.

Compounds differ from molecules in that their formation results from the combination of two or more atoms from different elements. The most well-known compound is water, formed from the association of two hydrogen atoms and one oxygen atom. Carbon dioxide is formed from the association of two oxygen atoms and a carbon atom.

2. Cells

Cells are the building blocks of all living things, either unicellular or multicellular. All metabolic activities are performed within cells. Metabolism, excretion, reproduction, respiration, irritability and growth are the basic functions that cells must perform.

Cells have vital roles in the survival of an organism despite their differences in structure and function. All are involved in different functions such as movement, support, energy etc. Thus, the unity of an organism is maintained by differentiated cells. Although cells differ in function and shape, they all possess an almost identical structure.

All are composed of a **plasma membrane** (or cell membrane), **cytoplasm** and **nucleus**. The plasma membrane is involved in the protection of the cell, providing shape, transport of materials and communication.

In plant cells, the plasma membrane is surrounded by a rigid cell wall. It is nonliving and is composed of cellulose molecules. Gates in the cell wall allow the passage of materials through the cells.

FIGURE: An animal cell

The cell wall provides shape for the plant cell and protects it against hazards.

The cytoplasm is the liquid portion of the cell located between the plasma membrane and the nucleus. It is composed of organelles, organic and inorganic molecules. The organic molecules of the cytoplasm are carbohydrates, proteins, fats and nucleic acids. The inorganic molecules are minerals, water, etc. The organelles of the cytoplasm are endoplasmic reticulum, Golgi apparatus, ribosomes etc.

The nucleus is the most vital component of the cell. It contains coded genetic information within its chromosomes. The cell can not survive if it is removed.

3. Tissues

A tissue is a group of cells specialized for a common purpose. Tissues are composed of cells and interstitial fluid involved in material exchange with the environment. An animal has four main tissues.

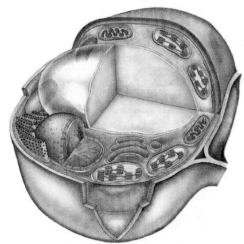

FIGURE: A plant cell

Animal tissues

Epithelial tissue: It is involved in secretion, protection and absorption and covers the body of most vertebrates. Additionally, it covers the inner surface of those organisms that possess a body cavity. Epithelial tissue also encapsulates the blood and lymph vessels and is classified according to its functions.

Connective tissue: It is generated from the mesoderm layer of the embryo. It is involved in the connection of all tissues and also supports the body. Connective tissue is composed of five main types;

Loose connective tissue (adipose, collagen fibers)

Fibrous connective tissue (tendons, ligaments)

Cartilage tissue

Bone tissue

Blood tissue

Muscular tissue: The body as a whole, as well as individual internal organs, move by means of muscular tissue. It is classified as skeletal, smooth or cardiac according to its structure. The skeletal muscles work in conjunction with bones to provide movement of the body. The smooth muscles contribute mostly to the structure of internal organs and to the movement of blood in the vessels. The cardiac muscles are responsible for the function of the heart.

Nervous tissue: It consists of neurons and glial cells and is involved in the transmission of impulses in the body.

- Epithelial tissue covers.
- Loose connective tissue supports.
- Fibrous connective tissue binds.
- Cartilage is flexible.
- Bone is rigid.
- Blood has a liquid matrix.
- Muscular tissue contracts.
- Nervous tissue conducts impulses.

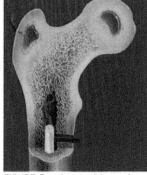

FIGURE: Bone is a special type of connective tissue.

Basic Concepts Of Anatomy And Physiology

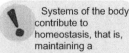

Systems of the body contribute to homeostasis, that is, maintaining a steady state of the internal environment.

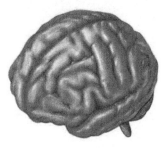

FIGURE: The brain is an organ of the nervous system.

WHO *ARE WE?*

- Human beings are highly organized.
- Human beings reproduce and grow, like other organisms.
- Unlike other living things, humans have a cultural heritage.
- Humans belong to the world of living things and are vertebrates.
- Humans can modify ecosystems.

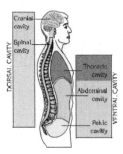

FIGURE: Body cavities.

4. Organs And Systems

An organ is the association of two or more tissues to perform a specific function. For instance, the lungs, liver, stomach, kidney and heart are organs that perform unique functions.

Groups of organs cooperate to perform functions, such as respiration, digestion, circulation, reproduction and excretion. These groups are known as systems, and examples include the circulatory system and the respiratory system.

The human body is an extremely complex structure which is dependent on the synchronization of the following systems.

Transport System	Respiratory System
Digestive System	Excretory System
Skeletal and Muscular System	Endocrine System
Nervous System	Sensory Reception

HUMAN PHYSIOLOGY

1. Body Cavities

The internal organs are positioned in the cavities of the abdomen, neck and head in both the anterior and posterior sections of the body.

The dorsal cavity consists of the skull and vertebral column. The skull houses the brain while the vertebral column houses the spinal cord. The lower portion of the ventral cavity is termed the lower abdominal cavity.

2. Fluids Of The Body

The human body consists of approximately 100 trillion cells which are combined to form fibers, projections and tissues. All spaces between them are filled by special fluids known as extracellular fluid or interstitial fluid.

The blood and its components fill the blood vessels and circulate around the body by the rhythmic contraction of the heart. In an adult body there are 42 liters of fluid. Approximately 25 liters is integrated into the cytoplasm of cells and is known as intracellular fluid, and 12 liters is involved in material exchange of cells and is known as interstitial fluid. The remaining 5 liters of fluid in the body is blood. Plasma accounts for 3 liters of this volume, the rest is blood cells.

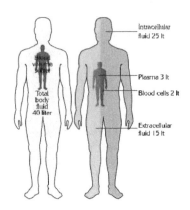

FIGURE: Fluids of the body.

The cells of the tissues are in constant contact with the fluids around them. They obtain oxygen, amino acids, glucose, hormones and other substances from this fluid.

They also excrete their metabolic wastes such as ammonia, urea, uric acid, CO_2 and hormones into this same fluid. The interstitial fluid is constantly replenished to ensure the survival of cells.

3. Membranes of the Body

Both the inner surface of the body cavity and the outer surface of the body are covered by mucosal, serosal and synovial membranes.

The mucosal membrane: The nasal cavity, mouth, urinary and reproductive canals, respiratory tract and alimentary tract are all covered by mucosa. These structures are directly or indirectly related to the external environment. The cells of the mucosal membrane secrete mucus to moisten these surfaces.

The serosal membrane: It is composed of loose connective tissue and is covered by epithelial cells. It is found in organs that have no connection with the outside environment. For instance, the pericardium of the heart is composed of a serosal membrane.

The synovial membrane: It covers the inner surface of joints such as those at the shoulder, pectoral girdle and knee. It is composed of connective and adipose tissue. It secretes a colorless, viscous fluid known as synovial fluid which is important for friction-free movement of joints.

4. Homeostasis

The internal balance of a healthy individual, known as homeostasis, remains constant despite the changes in the external environment. This is due to a high level of coordination between the structure and function of organs. This balance is maintained by both the nervous and endocrine systems.

The main factors involved in maintenance of homeostasis are blood pressure, body temperature, respiration and heart rate.

In a healthy individual, the chemical balance between the outer and inner environment is perpetuated by replenishment of interstitial fluid surrounding the cells.

The interstitial fluid is kept constant by the blood pressure, body temperature and the supply of sufficient materials to the cells.

All components of the human body therefore, have important roles in the maintenance of homeostasis. For instance, the kidneys filter the blood and separate soluble wastes and excess water for excretion.

Simultaneously, the digestive system is involved in hydrolysis of foods and their absorption into the blood. They are then distributed to all cells by the circulatory system.

The Regulation of Homeostasis

Homeostasis is regulated by the activities of the endocrine and nervous system. An individual will not suffer from any disorder as long as homeostasis is

 According to their cell structures, organisms are classified as prokaryotes (simple-celled) or eukaryotes (complex-celled). Bacteria are prokaryotes. Protista, fungi, plants and animals are eukaryotes.

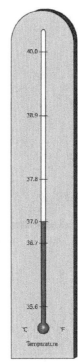

FIGURE: Body temperature at homeostasis.

 The temperature of the human body is maintained at about 37°C by the action of a regulator center in the hypothalamus, which is a part of the brain.

> Homeostasis of internal conditions is a self regulatory mechanism that usally results in slight changes above or below the balanced level.

maintained. If this balance is disrupted, the body may restore the balance and recover from the disorder. It may achieve this itself or by external medical intervention. Death is unavoidable if homeostasis is disrupted for long periods.

The first step in homeostatic regulation is the functioning of a feedback system in which the body attempts to reduce the adverse effects of the disorder. For instance, if an individual has a fever, the sweat glands are activated to reduce body temperature by perspiration. In cold conditions, the muscles under the skin contract and relax rapidly in a process known as shivering. The body temperature rises due to the heat produced by these movements. In this way, the body temperature is kept constant despite fluctuations in environmental temperature.

The negative feedback system is a bidirectional process. It either promotes or inhibits the secretion of hormones. For instance, low blood pressure is increased and returned to normal by a negative feedback system. A similar system reduces blood pressure if it increases above the normal level.

> There must be a certain balance between molecules, cells, tissues, organs, and systems of an organism to sustain homeostasis.

The positive feedback system differs completely from the negative feedback system. If a positive feedback system were activated when the blood glucose level decreased, it would cause a further decrease in the glucose level. The positive feedback system reinforces a condition instead of regulating it. Positive feedback systems are rare in the human body since many such systems would make homeostatic control impossible.

READ ME *STRES*

All systems of the body cooperate in the maintenance of homeostasis. Alterations both within the body and in the environment influence homeostasis. The factors that influence homeostasis are termed stressors, while the situation that they produce is termed stress. The stressor may have a physical nature, such as temperature or noise. It may be chemical, such as a hormone or food. It may also be pathological, namely bacterial or viral, or physiological due to tumors or abnormal functioning. A stressor may also be of psychological origin.

The immune system of an individual weakens if he can not manage the stress in his environment. Those who have personal problems, such as being unemployed, have an increased risk of heart disease.

Research has indicated that all diseases are related to some extent with stress. It can influence the progression of cancer, coronary diseases, lung disorders and psychological disorders both directly and indirectly. It may result in death if it can not be controlled.

CIRCULATORY SYSTEM

LYMPHATIC SYSTEM

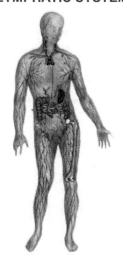

RESPIRATORY SYSTEM

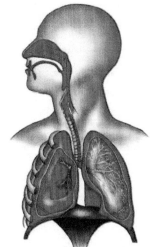

The human transport system consists of the heart, blood and vessels and is involved in the transport of nutrients, gases and wastes in the blood.

The transportation system of the body consists of only one organ, the heart. It has four chambers or cavities, and pumps blood through the body from before birth and continues to do so, never stopping or tiring, until the moment of death.

The heart is made up of involuntary cardiac muscle, but it is modified in structure as compared to all other involuntary muscle. After all, the heart beats an average of 72 times per minute every minute of your life.

The chamber of the heart that contracts most powerfully is the lower left chamber. This chamber is known as the left ventricle. When the muscles of the left ventricle contract, blood is forced in a mighty surge or pulse, out of a great artery that branches into all parts of the body. Arteries are the blood vessels that carry blood away from the heart.

The lymphatic system consists of glands, lymph nodes and lymph vessels. The lacteals are part of a somewhat separate transportation system of the body called the lymphatic system. The fluid inside the tubes of the lymphatic system is called lymph. Most of the waste products of the cells is picked up by the lymphatic system. These wastes are finally poured back into the blood stream in the same way that fatty acids are.

Not all of the lymph is in the lymph vessels. Every living cell of the body is bathed in a constant supply of lymph. It is the clear, watery substance that fills a blister, for example.

Lymph is actually just the liquid part of the blood. When food and oxygen diffuse out of the capillaries and into the cells, they become dissolved materials in this liquid part of the blood. Once out of the blood stream, the liquid is called lymph. Not all of the lymph diffuses back into the capillaries again, a considerable amount of it, filled with waste materials, diffuses into the lymphatic system and is carried back to the blood stream.

Every living thing must have a constant supply of materials in order to live. Some foods can be stored within the body of an animal or a plant. However, no living thing has ever developed a means of storing supplies of oxygen. Every living cell must be constantly provided with a supply of oxygen so that life will not cease. Thus, a continuous supply of oxygen must always be available.

The word respiration does not refer solely to breathing. This term means getting air into the lungs (inhalation) and removing the excess carbon dioxide that has been returned from the body to the lungs (exhalation). Respiration also includes the oxidation of food by living cells as well the transportation of oxygen to the cells and the removal of carbon dioxide, the result of cellular respiration. The term respiration includes the activity inside the cells, as well as the transport of oxygen and carbon dioxide between the cells and the lungs. It also includes the exchange of gases between the blood and the lungs, and the breathing process.

EXCRETORY SYSTEM

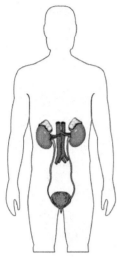

It consists of a pair of kidneys, ureters and the urinary bladder.

This system removes cell wastes from the blood stream and eliminates them from the body. Among the wastes removed from the blood stream are carbon dioxide, water, certain nitrogen compounds such as urea, and inorganic salts, mainly sodium chloride.

The lungs excrete carbon dioxide and water. Some water, salts, and a little urea are also removed from the body by the sweat glands. The principal organs of the excretory system, however, are the kidneys.

The kidneys excrete urine, composed mainly of nitrogenous wastes and various inorganic salts dissolved in water.

DIGESTIVE SYSTEM

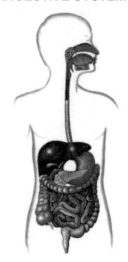

Ingested food must not only be broken up mechanically, it must also be changed to a soluble form, one that is water soluble. The body fluids carry the dissolved food by means of the blood stream throughout the body to the cells .

Food is broken down into molecules that dissolve in the body fluids. However, even after this occurs, it does not necessarily mean that these molecules can be absorbed by the cells.

The entire lining of the digestive tract is an unbroken membrane of epithelial tissue.

Furthermore, the outside structure of every cell in the body is an unbroken membrane. Therefore, for the molecules to enter the cells, they must be able to pass through these membranes.

There are, however, many molecules too large to pass through these membranes. The body overcomes this problem by the breakdown of large molecules into small molecules that are able to pass through the membranes.

SKELETAL SYSTEM

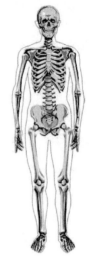

The backbone supports the entire body and must be resistant to all kinds of shocks during physical activity. In addition, it houses the spinal cord, a thick bundle of nerves running from the brain down through a hollow canal in the center of the backbone. Nerve tissue is extremely soft and delicate, and therefore easily damaged.

The spinal cord of nerve tissue controls many body activities. It carries nerve impulses from the brain to all parts of the body, and it carries other impulses, such as pain and temperature, from the body to the brain.

Therefore the backbone, which consists of 33 vertebrae, is a vital structure. Not only does it give support, it also protects the delicate spinal cord from injury.

MUSCULAR SYSTEM

Most of the muscles in the body are arranged in pairs. One of each pair bends a joint and the other moves the bones in the opposite direction. These pairs of flexors and extensors are attached to the bones by strong tendons.

Those at one end of a muscle are attached to one bone. The tendons at the other end of the muscle are usually attached to another bone. The two bones usually articulate at a joint.

Animals with an exoskeleton also have their muscles arranged in pairs. These muscles are arranged like those of a human arm. The muscles however, are attached to the exoskeleton rather than to bones.

Muscles such as flexors and extensors, are types of muscles known as voluntary muscles. That is, they are controlled by conscious will, at least in part. Most voluntary muscles consist of striated muscle cells.

ENDOCRINE SYSTEM

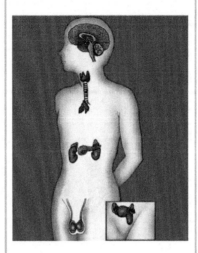

The endocrine system is composed of endocrine glands. The secretions of these glands are known as hormones, transported to their targets by diffusion or in the blood. Once at the target organ they regulate its function.

Generally hormones have a pronounced effect on metabolic functions; development, production, the level of glucose in the blood and on the concentration of minerals and water at specific levels. They also affect the permeability of the cell plasma membrane.

Hormones are extremely effective chemical substances due to their great potency. For this reason, they are found in small amounts in the blood and the urine.

NERVOUS SYSTEM

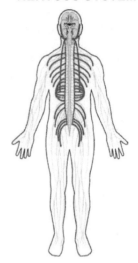

The nervous system is the communication system for the billions of cells that make up the body. Most types of animals have nervous systems which coordinate the activities of the cells that make up their bodies. Certain exceptions are the single-celled organisms.

Even they are thought by some to have types of fibers that provide communication between different parts of the body, and thus coordinate their activities.

Nerve cells are highly specialized. They stimulate muscle cells to contract and gland cells to produce secretions.

However, they cannot obtain food, digest it or expel their own waste. They are dependent on other cells, equally specialized to perform these functions for them.

Basic Concepts Of Anatomy And Physiology

Nervous system receives stimuli interprets them and prepares a proper answer to regulate harmony within the body.

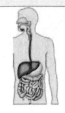

Digestive system change macromers into monomers in order to be taken into blood and then used by body cells.

Nitrogenous wastes, excess salts andminerals are removed from the body by means of excretory system.

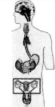

Endocrine system regulates body activities by using chemicals called hormones.

Circulation of needed substances and waste products within the body is provided by circulatory system.

O2 is needed for body cells to be used in cellular respiration and ATP

production. CO2 which is produced as a waste product of metabolism mustbe sent out. These two important activities are done by organs of respiratory system.

Locomotion system provides movement, support and protection.

Muscular and skeletal systems together from this system.

Nervous System

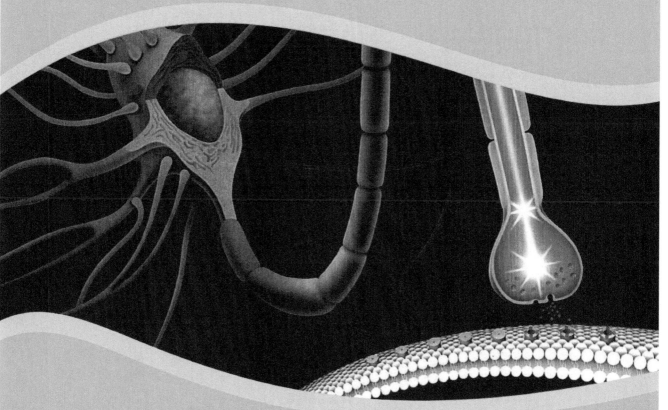

chapter 2

THE NERVOUS SYSTEM

The internal balance of an organism must be maintained in order for it to survive. This balance is known as homeostasis. An organism requires systems to maintain homeostasis and the normal functions of the body.

There are two main regulatory systems in the body: hormonal (chemical) and neural control systems.

The hormonal control system is found in both animals and plants. A neural control system, however, is found only in animals.

There are other ways besides the nervous system in which animal cells can communicate with each other.

One of the simplest ones uses cell junctions which allow the passage of ions and some molecules between nearby cells, but not ones that are far from each other.

Neurons, special cells of the nervous system, transmit information rapidly over long distances in the animal body.

The nervous system coordinates and regulates body functions. The five senses provide information on changes in the external environment.

Nervous connections to all organs of the body provide information on changes in the internal environment. The information is evaluated and a suitable response is elicited by effector organs such as muscles or endocrine glands. The nervous system connects the receptors to the effectors, evaluates messages and transmits a suitable response.

The stages in the function of the nervous system consist of reception, transmission, interpretation and response.

All these stages must obviously be performed within the same cell in unicellular organisms, whereas in multicellular organisms, the separate stages are performed in different structures within a system.

Even the most primitive organism shows some detection of a stimulus and a response to it.

Figure: Both external and internal stimuli are transmitted and interpreted by the CNS. A suitable response is then generated.

The nervous system is the ultimate coordinator of homeostasis. Nerves bring information to the brain and spinal cord from receptors that respond to changes both inside and outside the body. Then, the nerves take the commands given by the brain and spinal cord to effectors, allowing the body to respond quickly to these changes.

1. Organization of the Nervous System

A. The Neuron

1. The structure of a neuron

A nerve cell, known as a neuron, consists of the following components:

dendrites   cell body   axon

a. Dendrites

They are short, thin, numerous projections extending from the cell body. They receive information from other neurons.

b. Cell body

It is the enlarged part of the neuron from which dendrites and axons project. It contains the nucleus of the cell.

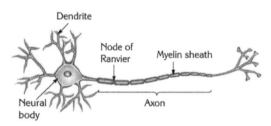

c. Axon

This structure resembles a dendrite in that it also projects from the cell body. In contrast to a dendrite, however, an axon is generally single, long and thick. Its length varies according to the location of the neuron, and may be in excess of 1 meter in length. The axon extending from the skull to the pelvis in a giraffe is about 3 meters long!

Figure: The structure of a typical neuron is composed of the above structures.

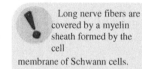

Long nerve fibers are covered by a myelin sheath formed by the cell membrane of Schwann cells.

Structure of an axon

Neurons are supported by specific cells known as neuroglia. Schwann cells and oligodendrocytes are two of the most important neuroglial cells, which surround the axons of many neurons.

During development, these cells wrap themselves around each axon several times to form the myelin layer. Myelin is an insulating material consisting of multiple layers of neuroglial membrane.

Transmission impulses are considerably more rapid in myelinated axons than in nonmyelinated axons.

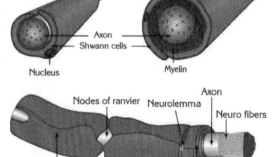

Axons are classified as myelinated and nonmyelinated according to their covering. Myelin is a thick, incomplete covering. Tiny gaps in it are known as nodes of Ranvier, which accelerate the speed of impulse transmission.

A nerve is a complex cord consisting of hundreds of axons united within connective tissue.

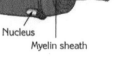

2. Classification of neurons

Neurons are classified according to their projections and functions.

a. Types of Neurons According to Their Projections

Neurons are classified according to the shape and number of their extensions.

Unipolar neuron: Only one short projection extends from the cell body.

Bipolar neuron: Two parallel projections extend from the cell body, one is the axon and the other is the dendrite.

Multipolar neuron: There are more than two projections. Dendrites are short, thin and numerous. Generally only one axon is present.

b. Types of Neurons According to Their Functions

Neurons are classified into three groups: sensory neurons, inter neurons and motor neurons

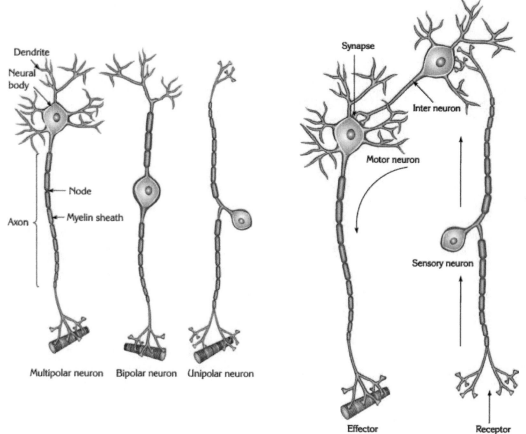

Figure: Neurons can be classified according to the number of projections. Figure: The transmission of an impulse occurs along neurons in the order as shown.

Sensory neurons transmit stimuli (Information gathered by receptors from sense organs or internal structures) from sensory organs such as the eye, ear, and skin to the CNS. Sensory neurons are usually unipolar or bipolar in structure.

Inter neurons are commonly found in the CNS. Their main function is interpretation of information. They are multipolar in structure.

Motor neurons transmit the impulses from the CNS to muscles or glands. Motor neurons are also multipolar.

3. Transmission of Impulses Through Neurons

Nerve impulses are electrical signals produced by the plasma membrane of a neuron.

The impulse generated at one end of the nerve is transmitted through the nerve fiber by electrical and chemical alterations. The impulse is generated if the voltage reaches a certain critical point, known as the **threshold level**. Any voltage lower than the threshold level fails to result in the generation of an impulse. This is known as the all-or-none law.

In nerve fibers, impulses are unidirectional, flowing in the same direction from dendrites to cell body to axon. The impulse is then transmitted to the dendrites of the next neuron, receptors of a muscle or a gland.

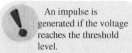
An impulse is generated if the voltage reaches the threshold level.
Any voltage lower than the threshold level fails to result in the generation of an impulse.

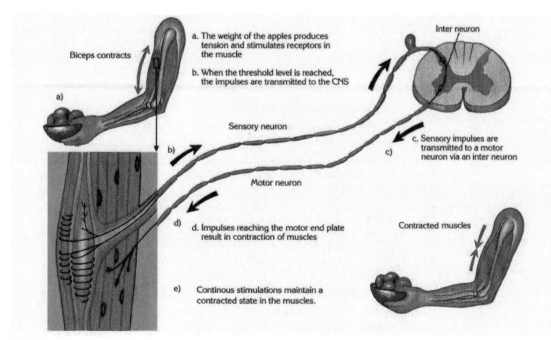

Biceps contracts

a. The weight of the apples produces tension and stimulates receptors in the muscle

b. When the threshold level is reached, the impulses are transmitted to the CNS

a)

b)

Sensory neuron

Inter neuron

c. Sensory impulses are transmitted to a motor neuron via an inter neuron

c)

Motor neuron

d)

d. Impulses reaching the motor end plate result in contraction of muscles

Contracted muscles

e)   Continous stimulations maintain a contracted state in the muscles.

Figure: The sequence of events that results in a response to a stimulus.

Nervous System

a. Mechanism of Impulse Transmission (Generation of Action Potential)

Phases of Impulse Transmission

Polarization

Depolarization (action potential) "The conduction of a nerve impulse"

Repolarization

Polarization: In a resting or unstimulated nerve, the outer portion of the axon is positively charged, while the inner portion is negatively charged (resting potential). This situation is called polarization.

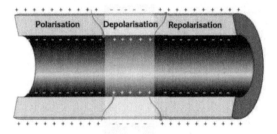

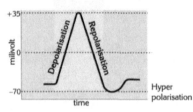

Figure: In a resting neuron, the potential difference across the membrane is -70 mV. During depolarization, this value increases to +35 mV. After the impulse has passed, the value returns to -70 mV.

Figure: Sodium-potassium pump. An unstimulated neuron is polarized with a net positive charge on the outer surface. Sodium ions are actively extruded. By the same mechanism, potassium ions are concentrated on the inner surface. As the impulse travels along the neuron, sodium ions flow in, depolarizing the membrane. After the impulse has passed, sodium ions are again pumped out of the cell, restoring the external positive charge.

What is the reason for polarity?

The existence of this polarity is correlated with a difference in ion distribution on either side of the axon. There is a higher concentration of sodium ions (Na^+) outside the axon and a higher concentration of potassium ions (K^+) inside the axon. The unequal distribution of these ions due to the action of the sodium-potassium pump is the main reason for polarity.

This pump is an active transport system in the cell membrane that pumps three sodium ions out for every two potassium ions taken into the axon. The pump is always working because the membrane is permeable to these ions. Since the membrane is more permeable to potassium ions, more positive ions are always found on the outside of the membrane. Another reason for polarity is that there are large, negatively charged proteins in the cytoplasm of the axon.

Depolarization: Besides the Na–K pump, there are two more pumps on the axomembrane. The sodium pump, which allows Na^+ to pass, and the potassium pump, which allows K^+ to pass.

A stimulus that reaches or exceeds the threshold opens the Na gates first. The

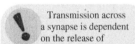

Transmission across a synapse is dependent on the release of

neurotransmitters, which diffuse across the synaptic cleft.

sudden entrance of the Na^+ makes a particular location inside the membrane positively charged. This is depolarization.

Repolarization: At a certain level, sodium gates are closed and potassium gates open. As a result, K^+ moves from the cytoplasm of the axon to the exterior of the axon.

As K^+ leave, the action potential once more return to the resting potential, and repolarization is achieved.

If an axon is unmyelinated, an action potential at one part stimulates an adjacent part of the axomembrane to produce an action potential.

In myelinated neurons, the action potential occurs at nodes of Ranvier and jumps from one node to the next. This type of transmission is called **saltatory conduction** (Latin saltare, to jump).

The velocity with which an action potential travels along an axon is greater if the diameter of the axon is large or if the axon is myelinated. In thin, unmyelinated axons, action potentials move about 1 m/s, and in thick, myelinated axons, the rate is more than 100 m/s.

B. Synapses

An impulse passing down an axon finally reaches the end of the axon and all of its branches. These branches may form junctions with muscle cells, with gland cells or with the dendrites of the next neuron. Such junctions are called **synapses**.

The neuron that carries the impulse to the synapse is called the **presynaptic cell,** while the cell located on the other end of the synapse is called the **postsynaptic cell**. Pre and postsynaptic cells don't make contact. There is a little gap between them called the **synaptic cleft**.

Through the synaptic cleft transmission is made by means of chemicals called **neurotransmitters**, which are carried in synaptic vesicles of the presynaptic cell.

When an impulse reaches the end of the axon, neurotransmitters are released into the synaptic cleft. They bind to receptors found on the membrane of the postsynaptic cell. There are many different types neurotransmitters.

Simply, a synapse is a junction between a neuron and another cell which can be a muscle cell, a gland cell or another neuron.

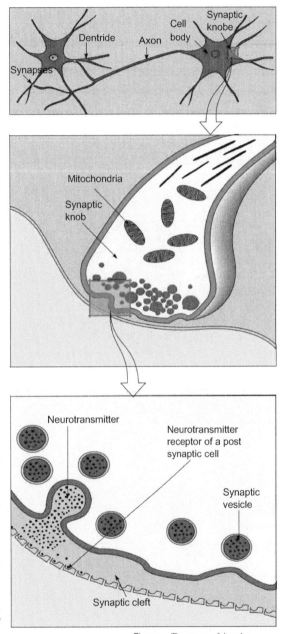

Figure: The stages of impulse transmission at a synapse.

Nervous System

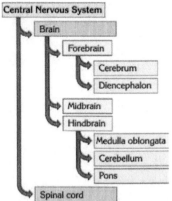

2. The Human Nervous System

The human nervous system consists of two main divisions:

Central nervous system (CNS)

Peripheral nervous system (PNS)

A. The Central Nervous System (CNS)

It consists of brain and spinal cord. These two structures are responsible for most of the information processing.

1. Brain

The brain, the most important organ of the central nervous system, is protected by the skull. The approximate weight of the brain is 1200-1350 g in males, 1000-1250 g in females, and its surface area is approximately 2000-2100 cm^2.

The brain weight of some famous people are as follows:

Trunjew -2012 g

Bismark -1807 g

The smallest known brain is approximately 369 g

The largest known brain is approximately 2850 g.

The central nervous system (CNS) is perfectly protected. There are three levels of protection:

1. Bony protection: The brain is protected by the skull, while the spinal cord is protected by the vertebral column. There are three layers of the skull covering the brain: **dura mater**, **arachnoid** and **pia mater**.

The dura mater is located directly beneath the periosteum and is separated from it by a space containing blood vessels and fat, known as the **epidural space**.

The arachnoid is found just under the dura mater and is separated by the **subdural space** which contains no cerebrospinal fluid. The arachnoid and pia mater are separated by the **subarachnoid space,** filled with cerebrospinal fluid, blood vessels and spinal roots.

2. Membrane protection: Both the brain and the spinal cord are protected by three main layers known as **meninges**. The meninges are composed of connective tissue.

3. Fluid protection: Cerebrospinal fluid functions as a shock absorber.

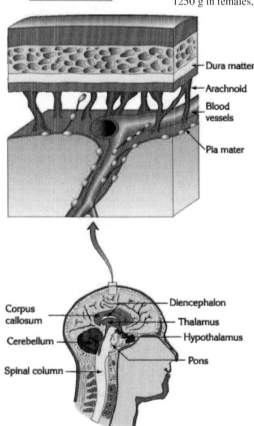

Figure: The brain is protected by the skull and meninges.

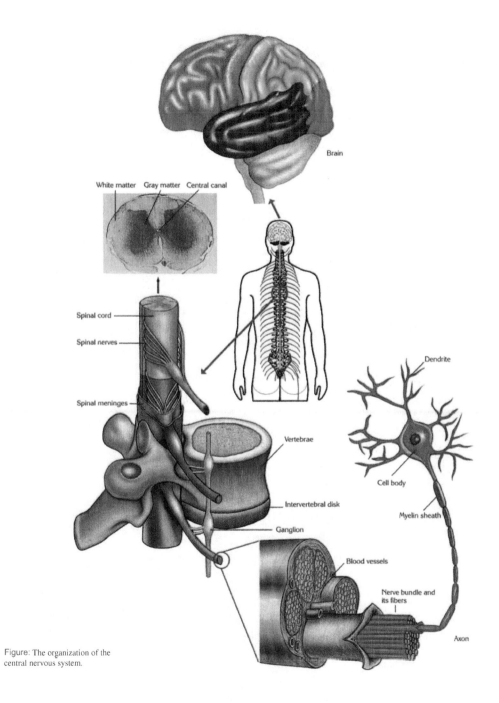

Brain

White matter Gray matter Central canal

Spinal cord

Spinal nerves

Spinal meninges

Vertebrae

Intervertebral disk

Ganglion

Blood vessels

Nerve bundle and
its fibers

Axon

Dendrite

Cell body

Myelin sheath

Figure: The organization of the
central nervous system.

The CNS, consisting of the brain and the spinal cord, receives sensory information and initiates motor control

What makes humans have much higher mental abilities than all other organisms?

An important condition affecting one in ten children is **dyslexia**. The extent to which sufferers are affected varies however. In this condition, the individual experiences problems in reading and writing since the order of letters perceived in certain words is confused. Children diagnosed as dyslexic need special help to overcome the problems that they encounter with reading and writing. It is thought that this condition develops during the second trimester of pregnancy due to an abnormal distribution of nerve cells within the brain.

The brain consists of three main parts:

Forebrain

Midbrain

Hindbrain

a. Forebrain:

It is the largest portion of the brain and is composed of two main divisions:

the cerebrum

the diencephalon

a. 1. Cerebrum

Cerebrum is the dominant (most active) part of the human brain. It is also the largest part of the brain. In longitudinal section, the brain is comprised of an outer cortex of **gray matter** surrounding all other structures of the brain.

This gray layer is mostly made up of dendrites and cell bodies of unmyelinated neurons, and has a thickness of 3-4 mm. Most of the activities of the cerebrum occur within this covering, also known as the **cerebral cortex**. The innermost layer is composed of axons of unmyelinated neurons and is termed **white matter**.

The cerebrum is divided into two cerebral hemispheres by a longitudinal fissure. Both of the cerebral hemispheres are interconnected by a single, thick bundle of nerve fibers called the **corpus callosum**.

The cerebral cortex of each hemisphere includes four main sections: **frontal lobe, parietal lobe, temporal lobe** and **occipital lobe**. There are three major activities of the cerebral cortex: motor activity (in the motor area), sensory activity (in the sensory area) and association activity (in the remainder of the brain that is not occupied by motor or sensory functions).

The motor area: It lies along the posterior (back) border of the frontal lobe. Each part of this area is related to the movement of a different part of the body.

The sensory area: It is located just behind the motor cortex, on the frontal part of the parietal lobe. Each part of this section receives inputs (stimuli) from sensory neurons connected to different parts of the body. Other sensory areas are located on the other lobes. For example, the auditory cortex is found within the temporal lobe. The visual cortex lies on the occipital lobe.

The remaining areas within the cerebral cortex are known as **association areas.** These are responsible for higher mental activities, such as planning.

The bigger the association area, the greater the ability of complex brain activities. In humans, 95% of all cerebral cortex is association area, while this ratio is only 5% in mice.

There are two lateral ventricles (spaces in the brain into which cerebrospinal fluid is secreted) in the brain, each within a hemisphere. These two ventricles are related to the third ventricle, which connects with the fourth by a narrow channel.

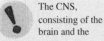

All our conscious life activities are managed by the different lobes of the cerebrum. This may be proved by an experiment in which the brain of a pigeon is removed. The animal is then observed to lose its memory and self-control. Furthermore, it shows no response to external stimuli. The pigeon can only walk if it is forced to or can only fly if thrown up into the air. Additionally, it feels no hunger and has to be force-fed.

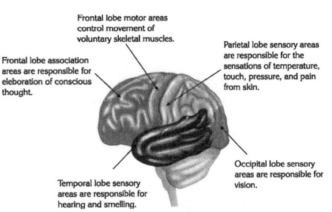

Frontal lobe motor areas control movement of voluntary skeletal muscles.

Parietal lobe sensory areas are responsible for the sensations of temperature, touch, pressure, and pain from skin.

Frontal lobe association areas are responsible for eleboration of conscious thought.

Occipital lobe sensory areas are responsible for vision.

Temporal lobe sensory areas are responsible for hearing and smelling.

Figure: The cerebral cortex.

As explained in the experiment, the conscious activities listed above are regulated by the left and right cerebral hemispheres. In right handed people, the left hemisphere is involved in speech, writing, calculation, comprehension of language and analytical thought.

The right hemisphere is involved in comprehension of simple language, general thought processes, conceptual, nonverbal ideas and appreciation of spatial relationship and music.

a. 2. Diencephalon

Also known as the midbrain, it includes the thalamus and hypothalamus. At the base of the cerebrum there are paired oval masses of gray matter called **thalamus** (inner chamber).

It functions as a relay station for most sensory information The messages received from the spinal cord and certain parts of the brain are sent to the appropriate areas of the cerebral cortex.

The hypothalamus is located directly beneath the thalamus and third ventricle. This vital structure is responsible for integration of visceral activities, such as regulation of body temperature (along with the limbic system), many emotional feelings including anger, pain, pleasure, hunger, thirst, and sexual desire.

It also controls the pituitary gland, which in turn regulates many of the endocrine glands. The hypothalamus maintains homeostasis of the body by integration with the autonomic nervous system.

It influences blood pressure, peristalsis and glandular secretions of the digestive system

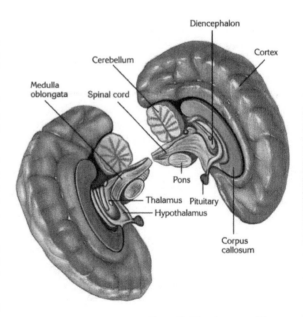

Diencephalon

Cortex

Cerebellum

Medulla oblongata

Spinal cord

Pons

Thalamus Pituitary

Hypothalamus

Corpus callosum

Figure: The bilateral structure of the human brain, showing its components.

Consciousness is the province of the cerebrum, the most developed part of the human brain.

The cerebrum is responsible for higher mental activities.

The midbrain, pons, and medulla together form the brainstem.

Tracts of nerve fibers in the brainstem bring messages to and from the spinal cord.

b. Midbrain

It is located between the diencephalon and the pons, and connects the forebrain, pons and cerebrum. It has special structures involved in the coordination of head and eye movement. They focus the image from the eyes, and also adjust the pupils according to stimuli. Furthermore, some of these structures indirectly affect the auditory cortex and hearing.

c. Hind brain

It is composed of the medulla oblongata, cerebellum and pons.

c. 1. Medulla Oblongata

It is like a projection of the spinal cord in the brain. It is located adjacent to the posterior of the pons. Its length is 3 cm and its weight is 8-10 g. The cerebral nerves cross over at the medulla oblongata and extend to the internal organs of the body. Thus, the left cerebral hemisphere controls the right side of the body, whereas the right hemisphere controls the left side of the body.

The medulla oblongata is composed of gray matter on the outside and white matter on the inside, as in the brain.

This structure is surrounded by meninges. It regulates the respiratory, circulatory and excretory activities, and many other involuntary visceral activities, such as the regulation of glucose metabolism by the liver, swallowing, sneezing, vomiting, vasodilation and vasoconstriction.

Midbrain, pons and medulla together form the **brain sytem**.

The association areas are responsible for higher mental activities such as planning. The bigger the association area the greater the ability of complex brain activities. In humans, 95% of all cerebral cortex is association area, while this ratio is only 5% in mice.

READ ME

The degree of branching of neurons and subsequent synapsing is low in very young children. Stimulation given to developing children increases the frequency of synapsing in the brain, leading to greater intelligence.

A single neuron has the capacity to synapse with up to 10,000 other neurons.

INTELLIGENCE

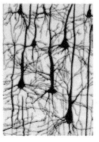

c. 2. Cerebellum

Below the cerebrum, in the rear of the cranium, is the cerebellum. It is located adjacent to the anterior of the medulla oblongata. It is connected to the medulla oblongata and midbrain. It is surrounded by the dura mater, arachnoid and pia mater, as in the brain. In transverse section, gray matter is located around the white matter.

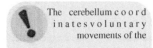

The cerebellum c o o r d i n a t e s v o l u n t a r y movements of the skeletal muscles.

It coordinates the movement of the voluntary muscles so that they work together. The act of picking up a piece of candy with the fingers and putting it in the mouth requires the coordination of many muscles, all working together. The control of the body's balance is also centered in the cerebellum.

The functions of the cerebellum

It is the center of locomotion and balance. Thus, the size of the cerebellum is proportional to the organism's capacity to move.

It coordinates the functions of muscles and is subsequently large in extremely mobile organisms such as birds. If the cerebellum of a dog is removed surgically, it loses the ability to walk. In the same way, a pigeon loses the ability to fly if the cerebellum is nonfunctional.

Any external stimuli received by sensory organs are carried to the cerebellum by sensory neurons.

The cerebellum generates impulses to respond to the situation. The balance of the body is maintained by the cerebellum and semicircular canals found within the ears. Some pressure receptors located at the base of the feet also help the cerebellum to keep the body in balance during movement.

c. 3. pons

It is a bridge-like structure between the midbrain, medulla oblongata, cerebral hemisphere, cerebellum and cerebrum. It controls certain respiratory functions.

2. The Spinal Cord

It is a brain extension. It is approximately 45-50 cm in length. The superior tip (upper part) is closely connected to the brain stem, whereas the anterior tip is a cone-shaped blind point. There is a tiny hole running through the spinal cord known as the vertebral canal which is formed by the spaces (foramen) of vertebrae. This bony case protects the spinal cord. It is also protected by meninges and cerebrospinal fluid, as mentioned before.

In cross section, the spinal cord consists of two distinct areas. The innermost segment is gray in color and resembles a butterfly in shape.

The gray matter is composed of cell bodies and is surrounded by a layer of white matter. The gray matter has two posterior columns and two anterior horns. From the anterior (frontal) horns motor neurons leave the spinal cord while stimuli brought by the sensory neurons enter the spinal cord from the posterior (back) horns.

The Limbic System
It lies just beneath the cerebral cortex and contains neural pathways that connect portions of the frontal lobes, the temporal lobes, the thalamus, and the hypothalamus.

By causing pleasant or unpleasant feelings about experiences, the limbic system influences how the individual will behave in the future.

The limbic system is absolutely essential to both short-term and long-term memory.

As a result it is possible to say that, "The limbic system is particularly involved in emotions, in memory and learning"

The part of the limbic system responsible for memory is called the hippocampus.

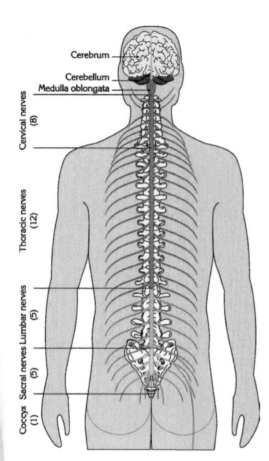

Cerebrum
Cerebellum
Medulla oblongata

Cervical nerves (8)

Thoracic nerves (12)

Lumbar nerves (5)

Sacral nerves (5)

Coccyx (1)

Figure-1. 23. : The spinal nerves radiate from the spinal cord, housed in the vertebral column.

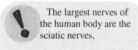

The largest nerves of the human body are the sciatic nerves,

which extend into both legs.

Functions:

The two basic functions are:

Transmission of impulses to the brain and tissues. The spinal cord carries information to the brain and, similarly, from the brain out to the body.

Facilitation of reflex actions. A reflex is an automatic response to nerve stimulation. No thinking takes place during reflex action. The pathway an impulse follows during reflex activity is called a **reflex arc**. Impulses travel along the sensory neurons to the spinal cord, which then sends messages out via the motor neurons without involving the brain.

The spinal nerves

The interneurons of the spinal cord interconnect the sensory and motor neurons. They constitute 31 pairs of nerves originating from the spinal cord. Each pair is known as a segment.

These 31 segments are classified according to the area of the body with which they communicate. Thus, 31 pairs of nerves originate from the spinal cord. Eight of them originate from the cervical region, 12 from the thoracic region, 5 from the lumbar region and 5 from the sacral region. One pair also originates from the coccyx. The diameter of the spinal nerve is directly proportional to the diameter of the target organ. The largest nerve of the human body, therefore, is the sciatic nerve which extends into each leg.

a. Reflexes:

A reflex is the transmission of impulses, generated by receptors, to the target, where a sudden response is generated. An example of a reflex arc can be explained as follows: When an individual touches a hot object, receptors in the skin sense it and an impulse is generated in the sensory neurons. The impulse is carried to the interneuron, which has a synaptic link, synapsing with both sensory neurons and motor neurons. The interneuron transmits the impulses to the motor neuron and finally to the effector organ.

The knee-jerk reflex is a simple (monosynaptic) reflex which includes only two neurons. In this reflex, a sensory neuron communicates directly with a motor neuron. The knee-jerk reflex is initiated by striking the patellar ligament located directly beneath the patella. The receptors located in the muscles attached to the patella by tendons are thus stimulated, generating an impulse in the sensory neuron. In the spinal cord, the sensory axon forms a synapse with the dendrite of the motor neuron. The impulse is carried to the high and the muscles respond by contracting. The reflex is completed as the lower leg moves upward. The brain is aware of this action at the knee since some impulses also pass simultaneously to the brain.

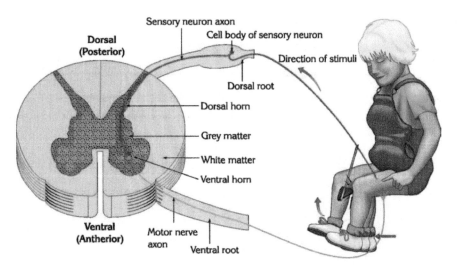

Figure: A simple reflex, such as the knee-jerk reaction, results from the activity of only two neurons, motor and sensory.

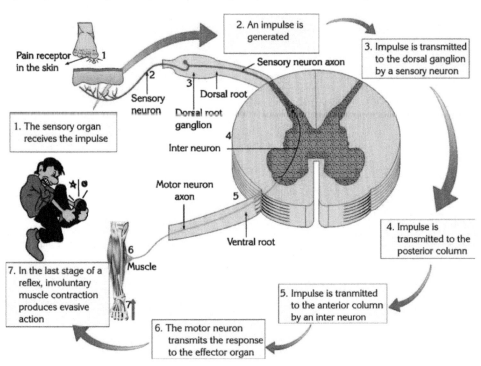

Figure: The events that form a reflex action.

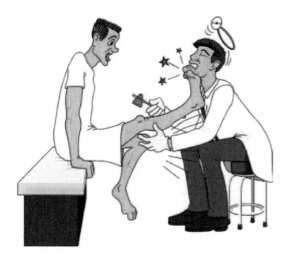

Most reflex arcs are more complex than the knee-jerk reflex. Three or more neurons are involved in a type of reflex known as withdrawal reflex. It occurs when a person unexpectedly touches something painful. An individual will raise his foot immediately after stepping on a nail, for example, and will then move away from the source of the problem to reduce the risk of further injury.

The **withdrawal reflex** is inherited, therefore it is not gained through experience. Most reflexes are involuntary and are observed in most organisms.

The activities of the spinal cord are regulated by the brain, which can override some reflex actions. For instance, it would be expected that a tight rope walker would lose his balance and fall if he steps on a sharp object, since this would be the body's natural response.

In reality, the brain intervenes to dominate the reflex. Thus, the tight rope walker can continue to walk on the high wire in spite of the pain.

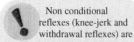 The reflex arc is a major functional unit of the nervous system. It allows us to react rapidly to external and internal stimuli.

Some reflexes are stored by the brain after the experience. These types of reflexes are termed **conditional reflexes**. The Russian physiologist Pavlov conducted some now famous experiments with dogs. In the first step, the dogs were fed on milk during their first year of life.

From then on, if they saw milk or, for example, meat, they salivated. They were then given meat on a few occasions. In the next step, Pavlov observed their behavior when they saw only meat. He noted that they began to salivate. These investigations indicated that the initial stimulus was due to observation.

Pavlov experimented further, observing the effect of irrelevant stimuli on his dogs. They became conditioned to such stimuli as a bell ringing or an electric shock. Such responses he termed classical conditioning.

He observed that the dogs secreted saliva in response to such stimuli. Initially, there was no response. However, if he rang a bell whenever the dogs were fed, they became conditioned to the sound of the bell and released saliva whenever they heard it ringing.

Non conditional reflexes (knee-jerk and withdrawal reflexes) are inherited, while conditional reflexes are nonhereditary and have a great importance in human physiology.

Simple reflexes and withdrawal reflexes are the basic responses of the body and no physiological difference exists between conditional and nonconditional reflexes. Conditioning of strong inherited reflexes presents great difficulties. Conditioning of weak temporary reflexes, however, is easy. Conditional reflexes are nonhereditary and have a great importance in human physiology.

The activity of a nerve is continuous. Nerves never become fatigued when sufficient food and oxygen is supplied.

Reflexes however, are susceptible to fatigue. The knee jerk response can not be continued indefinitely. Withdrawal reflex arcs in particular are easily fatigued.

B. The Peripheral Nervous System (PNS)

It consists of nerves that connect the brain and spinal cord to the other organs of the body. It comprises sensory nerves, motor nerves and complex nerves. The peripheral nerves originate from the brain and spinal cord. The PNS is composed of 12 pairs of cranial (coming from the brain) and 31 pairs of spinal (extending from the spinal cord) nerves

The peripheral nervous system is divided into two systems according to its function and mechanism:

somatic nervous system

autonomic nervous system

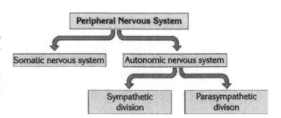

1. The Somatic Nervous System (SNS)

It consists of both motor and sensory nerves. The cell body of these nerves is anchored in the brain and spinal cord, and their axons communicate directly with the skeletal muscles.

The behavioral and physical activities of the brain, such as running, writing, painting and singing, are coordinated by this system. The somatic nervous system controls and coordinates the voluntary actions of the body.

2. The Autonomic Nervous System (ANS)

It regulates the involuntary activities of internal organs and is basically involved in the maintenance of homeostasis. It plays an important role in the regulation of respiration, excretion, circulation and digestion as well as nutrition, reproduction and adaption to changes in environmental conditions.

The autonomic nervous system consists of sympathetic and parasympathetic divisions which function antagonistically. The effector organs of the ANS are smooth muscle fibres which form an integral part of the internal organs, cardiac muscle and glands.

The ANS is classified as follows:

Sympathetic division

Parasympathetic division

The sympathetic division initiates **fight or flight responses,** while the parasympathetic division prepares the body for relaxation and digestion of food.

a. The Sympathetic Division:

Axons leave the spinal cord at the thoracic and abdominal levels and travel a short distance to a chain of ganglia just outside the spinal cord. There they synapse with neurons whose axons innervate the internal organs

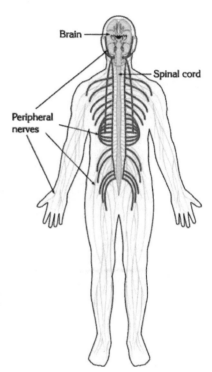

Figure: Position of autonomic and somatic nerves of the body

 The sympathetic system brings about those responses we

associate with "fight or flight".

The parasympathetic system brings about the responses we associate with a relaxed state.

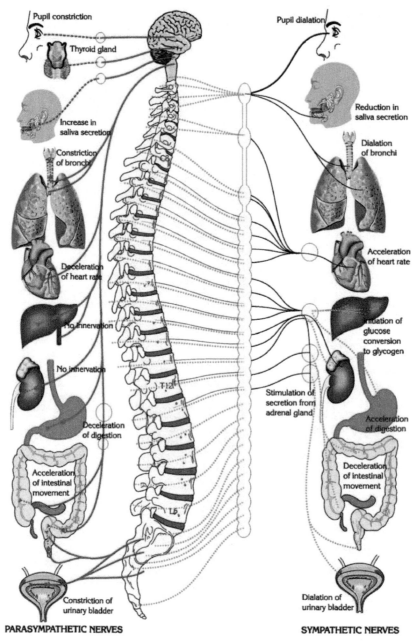

Pupil constriction

Thyroid gland

Pupil dialation

Reduction in saliva secretion

Increase in saliva secretion

Constriction of bronchi

Dialation of bronchi

Deceleration of heart rate

Acceleration of heart rate

No innervation

Initiation of glucose conversion to glycogen

No innervation

Stimulation of secretion from adrenal gland

Deceleration of digestion

Acceleration of digestion

Acceleration of intestinal movement

Deceleration of intestinal movement

Constriction of urinary bladder

Dialation of urinary bladder

C8

T12

L5

PARASYMPATHETIC NERVES

SYMPATHETIC NERVES

Figure: Effector organs of sympathetic and parasympathetic nerves.

ROENCEPHALOGRAM

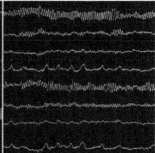

Electroencephalogram, or EEG, is an observable form of the weak electrical activity of the cerebral cortex.

Most of the natural electrical impulses of the brain can not be detected due to the presence of the skull. However, the many weak impulses that pass through the skull can be detected by electrodes.

The EEG in the figure is of a relaxed, conscious individual. In a calm but wakeful state, a-waves form regular patterns which are lost when the eyes are opened to the many stimuli in the environment.

During thinking, these waves deepen, indicating increased electrical activity. It is only possible, however, to detect that the process of thinking is occurring, not the nature of the thought.

The EEG is an important diagnostic tool in the treatment of brain disorders and the effects of drugs.

The sympathetic system is especially important during emergency. It accelerates the heartbeat and dilates bronchi. On the other hand, it inhibits the digestive tract (When we are under attack digestion is not an immediate necessity). The primary neurotransmitter in SNS is norepinephrine. Its structure is like that of epinephrine (adrenaline), a hormone produced by the adrenal medulla.

b. Parasympathetic division:

Axons extend from the brain and sacral (lower back) region of the spinal cord to the various internal organs (glands, smooth and cardiac muscles). Sometimes it is called the "housekeeper system". It brings about the responses we associate with a relaxed state. For example, it causes the pupil of the eye to contract, promotes digestion of food and retards the heartbeat. The neurotransmitter that is used by the parasympathetic division is acetylcholine (ACh)

Diseases and Disorders of the Nervous System

Epilepsy

Epilepsy is a condition resulting from disruptions in the normal electrical activity of the brain.

During wakefulness and sleep, the brain produces electrical waves. During an epileptic seizure, the electrical activity increases dramatically. Two forms of epilepsy exist:

Minor epilepsy: This form usually starts in childhood. An attack may be almost unnoticeable. The sufferer becomes distant and may behave strangely.

Major epilepsy is violent and unexpected. The sufferer loses consciousness, becomes rigid and may stop breathing.

 Some drugs resemble neurotransmitters in structure and function.

For example: Dopamine is a natural neurotransmitter in the brain, while cocaine is a drug that shows the same effects as dopamine.

The limbic system is the main part of the brain affected by drug abuse.

When drugs are used, mood and emotions can be altered.

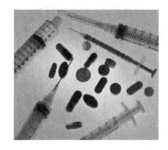

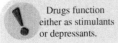
Drugs function either as stimulants or depressants.

Alcohol is the most abused drug all over the world.

It influences neurotransmitters and affects impulse transmission. Alcohol is metabolized in the liver. When there is alcohol in the body, the liver can not do its cellular respiration normally. Fat metabolism stops and the liver becomes fatty, the first stage of liver deterioration, beginning after only a single bout of heavy drinking. If the person continues to drink alcohol, it may cause cirrhosis (liver destruction) and even death.

Figure: Alcoholics usually behave unconsciously.

In the bodies of people who drink alcohol; the immune system is depressed and the chance of having colon (intestine), liver, lung, pancreas and stomach cancer increases.

The body then starts to jerk uncontrollably, the sufferer remaining unconscious. An attack may last for up to a few minutes. It is important not to try to intervene while an epileptic seizure is in progress. However, all articles that might injure the epileptic should be removed. Those suffering from epilepsy are prescribed drugs to prevent the onset of an attack. They on no account should be allowed to drive.

Meningitis

Meningitis is the inflammation of the meninges or lining of the brain. It may result from either bacterial or viral infection. Bacterial meningitis is the most life threatening form of meningitis but is quite rare. Unless bacterial meningitis is diagnosed promptly, it can prove fatal.

Bacterial Meningitis

It is caused by pneumococcal or meningococcal bacteria. This type of meningitis is treatable with antibiotics if a diagnosis is made early enough. It is a rare condition and some of those affected never recover, or develop brain damage or deafness.

Viral Meningitis

It is caused by the virus Hemophilus influenza type b. This form of meningitis is more common and is less life threatening. In recent days, a vaccine against viral meningitis has been produced.

The symptoms of both forms are similar and it is vital to hospitalize the patient as early as possible if meningitis is suspected.

The following symptoms may develop over a few days or hours: headache, fever, vomiting, stiffness of the neck and joints, dislike of bright lights, drowsiness. In addition, a purple rash of spots can develop under the skin, anywhere on the body.

If you don't smoke, do not smoke. If you smoke, it is never too late to quit.

Nicotine causes both physiological and psychological dependence. That is why it is quite difficult to give up smoking. Nicotine functions as acetylcholine (neurotransmitter) and increases skeletal muscle activity, heart

rate and blood pressure. Nicotine increases digestive tract mobility, which is why smokers want to smoke after eating.

SELF CHECK

NERVOUS SYSTEM

A. Key Terms

Neuron	Synapse
Nodes of Ranvier	Action potential
Homeostasis	Cerebellum
Peripheral N.S.	Limbic System
Cerebrospinal fluid	Meninges
Unipolar neuron	Motor neuron
Cerebral hemispheres	Corpus callosum

B. Review Questions

1. Explain the properties of the nervous system.

2. Classify neurons according to their projections and explain their structures.

3. Classify neurons according to their functions.

4. Draw diagrams and explain action potential, polarization, depolarization and repolarization.

5. Draw diagrams and explain the stages of synapses.

6. Explain the functions of synapses.

7. Explain the structure and function of the meninges of the brain.

8. List the components of a reflex arc.

9. Explain the events of a reflex action, from a receptor to an effector.

10. Explain the function of cerebrospinal fluid.

11. Compare conditional and nonconditional reflexes.

12. Explain the function of the brain in withdrawal reflexes.

13. Explain the properties of the somatic nervous system.

14. Explain the structure, components and functions of the autonomic nervous system.

15. Compare the somatic and autonomic nervous systems.

C. True or False

1. Dendrites are long extensions of cell bodies that receive information from other neurons.

2. Connection sites of neurons and effectors are called neurotransmitters.

3. The part of the PNS that connects the CNS to the part of the body which works voluntarily, is the somatic nervous system.

4. The end of an axon and the membrane of the effector are not directly connected. The space between them is called the cell body.

5. Unlike the brain, the inner portion of the spinal cord is gray in color, and called gray matter.

D. Matching

a. Axon   ( ) Part of the brain that provides higher mental functions, consisting of two hemispheres.

b. Effector   ( ) Schwann plasma membranes that cover long neuron axons.

c. Midbrain   ( ) Fiber of a neuron that conducts nerve impulses away from the cell body.

d. Myelin sheath   ( ) The part of the brain that regulates pupil diameter and position of the eye balls.

e. Cerebrum   ( ) Structure, such as a muscle or a gland, that allows an organism to respond to internal and external stimuli.

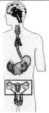

Hypothalamus, as a part of brain, controls endocrine system activities. Sex hormones have role in brain development.

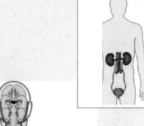

Muscles, permit urination, are controlled by the brain.

Kidneys maintain concentration of some minerals, needed for nervous system activity, at a certain level.

Skeletal System

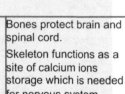

Bones protect brain and spinal cord.

Skeleton functions as a site of calcium ions storage which is needed for nervous system activity.

Nervous system coordinates skeletal functions

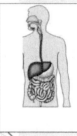

Peristaltic and certain digestive movements done by smooth muscles are controlled by nervous system. Digestive system provides nutrients for the cells of the nervous system.

Muscular System

Brain controls muscular activities.

Nerves of PNS are protected by muscle layers.

Rate of respiration is controlled by medulla oblangata and pons. Respiratory system provides oxygen to the nervous system and remove carbondioxide from it.

Heartbeat rate and vascular contriction is regulated by nervous system.

Circulatory system delivers oxygen and nutrients to the nervous system and removes wastes from its cells

BIOLOGY
HUMAN

Senses

chapter 3

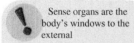

Sense organs are the body's windows to the external environment.

Receptors are specialized to receive stimuli coming from internal or external environments.

Photoreceptors detect light.

Pain receptors detect chemicals released by damaged cells.

Proprioceptors maintain the body posture.

Thermoreceptors are stimulated by changes in temperature.

Chemoreceptors are sensitive to dissolved chemical substances.

SENSORY RECEPTION

An organism can only survive if it is able to detect changes both in its environment and in its body. The physical and chemical changes in the environment stimulate free nerve endings or receptors connected to the nerves. These external stimuli cause the generation of impulses. The nerves transmit impulses to some centers, and the response is transmitted to the effector organ by other nerves. Organisms may respond to stimuli by effector organs such as muscles and glands.

All stimuli cause alterations in the membrane potential of receptors. If this potential difference reaches the threshold level, an impulse is generated. These impulses stimulate different regions of the central nervous system. This results in different interpretations in the brain even though the initial stimulus and pathway of transmission is the same. The CNS is informed of stimuli in the form of light, temperature, taste, smell etc.

The coordination of receptors, intermediate structures and the CNS is important for the successful transmission of an impulse. An impulse can not be generated if the receptor is nonfunctional. Neither can it be transmitted if the nerve fibers are severed, as in paralysis. The impulse can not be interpreted if the CNS is deactivated.

Sense organs are known as the "windows of the brain" because they detect and send nerve impulses (any changes in the environment and internal portions of the body) to the central nervous system. Information reaching the cerebrum of the brain results in conscious sensation.

RECEPTORS AND SENSATION

Receptors are structures specialized to receive certain enviromental stimuli and change them into nerve impulses.

According to stimuli, receptors in humans can be classified into six types:

Photoreceptors detect light. Eyes have this kind of receptors.

Pain receptors are naked nerve endings that respond to chemicals released by damaged tissues or excess stimuli of pressure or heat.

Proprioceptors sense the degree of muscle contraction, the stretch of the tendons (structures that connect muscles to other muscles or to bones), and the movement of ligaments (strong connections between bones). Information sent to the central nervous system by these receptors is used to maintain the body's posture.

Thermoreceptors are stimulated by changes in temperature. Those that respond when temperature rises are called heat receptors, and those that respond when temperature decreases are known as cold receptors. There are internal thermoreceptors in the hypothalamus and surface thermoreceptors in the skin.

Chemoreceptors are sensitive to dissolved chemical substances. The sense of taste and smell are well-known types of chemoreception. Various internal organs also have chemoreceptors (one monitors the pH level of the blood).

Mechanoreceptors are stimulated by mechanical forces, which are most often different types of pressure. Skin and ears are sense organs that have this kind of receptor.

THE HUMAN EYE

The eyes of all vertebrate organisms are structurally similar to each other and resemble a simple camera in function.

A vertebrate eye consists of two main structures: an eye sphere and accessory structures.

The eye sphere is 2.5 cm in diameter and weighs 10 g. It functions as a camera and is composed of three layers.

1. The structure of the eye sphere

It consists of the following components:

sclera   choroid   retina

a. Sclera

It functions as a supportive structure and is composed of fibrous connective tissue. The part of the sclera in front of the eye is transparent and is called the **cornea**. Because the cornea is rounded, it not only allows light to enter the eye but bends it as well (refraction). Therefore, the cornea is the first part of the eye that focuses the light.

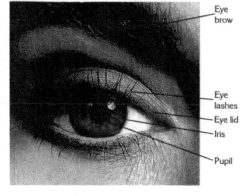

Eye brow

Eye lashes

Eye lid

Iris

Pupil

A chamber behind the cornea is filled with a watery fluid called the aqueous humor. This fluid nourishes both the cornea and the lens, since neither structure has a supply of blood vessels (aqueous: water; humor: body fluid). A small amount of this fluid is produced each day, and old fluid enters the blood stream. If this drainage is blocked, it causes **glaucoma**. If it is not treated it can cause blindness. In addition to aqueous humor, a clear, gelatinous material, the vitrous humor (vitrous: glass), fills the posterior cavity behind the lens. It makes a pressure within the eyeball that maintains the eyeball's shape.

The lens is located just behind the aqueous humor and plays an important role in focusing the light onto the proper part of the eye, the retina, where photoreceptors are found

b. Choroid

This layer is rich in blood vessels and pigments and prevents the internal reflection of light due to its dark brown color. The blood vessels of the choroid supply nutrients to the retina. The ciliary body and the iris are located in this layer.

Ciliary body (corpus ciliare): It is a smooth muscular structure located directly beneath the sclera.

Muscles exert an influence on the lens via these fibers, and enable adjustments to be made in order for an image to be formed on the fovea (the part of the retina where many of the photoreceptors are located). There are approximately 70 muscular fibers around the lens.

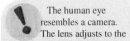

The human eye resembles a camera. The lens adjusts to the distance of the object. The iris functions like a diaphragm and regulates the amount of light entering the eye through an opening called the pupil. Instead of a light sensitive film, the retina includes light sensitive receptors, the cones and the rods.

SENSES

39

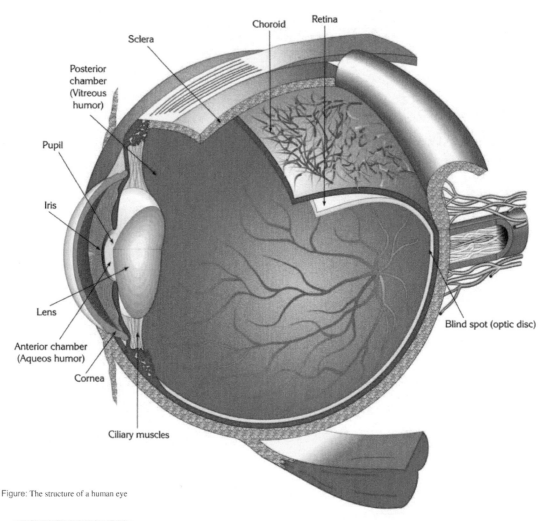

Sclera

Choroid Retina

Posterior
chamber
(Vitreous
humor)

Pupil

Iris

Lens

Anterior chamber
(Aqueos humor)

Cornea

Blind spot (optic disc)

Ciliary muscles

Figure: The structure of a human eye

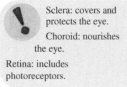

Sclera: covers and
protects the eye.
Choroid: nourishes
the eye.
Retina: includes
photoreceptors.

Iris: It is located between the muscular layer and the retina and is composed of smooth muscle. Pigments are located in this structure, giving each individual an eye color, such as brown, blue or green. In albinos, the light passes directly from the cornea to the retina due to an absence of pigment. If pigment were present in the iris, it would force light to enter the eye through the eye lens.

The amount of light entering is regulated by the contraction and relaxation of the smooth muscles of the iris, which function like the diaphragm of a camera. The hole at the center of the iris through which light enters is termed the **pupil**.

c. Retina

The innermost layer of the eye sphere, the retina includes two main types of photoreceptors, rod cells and cone cells. The retina is made of three main functional layers. The rod and cone layer is nearest to the choroid. Next comes the bipolar cell layer. The innermost part is the ganglionic cell layer, whose axons are part of the optic nerve. Because rod and cone receptor cells are located on the back layer, light must reach them. The retina includes approximately 125 million rod and 7 million cone photoreceptor cells. The rod cells contain the light sensitive molecule **rhodopsin**. The precursor of rhodopsin is vitamin A, and is involved in vision under dark conditions (dim light). Any lack of rhodopsin results in difficulties in vision at night. Thus, vitamin-A deficiency results in the condition known as night blindness. Cone cells function in bright light and detect colors. The area where the greatest degree of detail and color detection is possible is known as the **fovea**. The fovea includes only cone cells. There are no rod or cone cells where the optic nerve passes through the retina, and vision here is impossible. This part of the retina is called the **blind spot**.

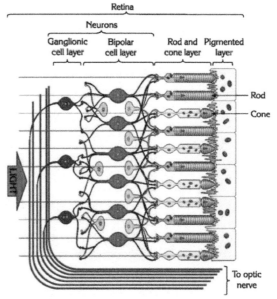

Figure: The structures and pathway involved in reception of light and its transmission.

2. External Structures of the Eye

Other associated structures of the eye are: the orbit of the skull, eye lids, eyelashes, eyebrows, conjunctiva, tear apparatus and muscles. All these structures contribute to the eye and its effective function

a. Orbit

The eye is located in a cavity called the orbit. It cushions the eye, protecting it from external hazards. Orifices (holes) in the wall of the orbit permit penetration of blood vessels and nerves into the eye sphere.

b. Eyelids

They protect the eyeball from harm: for example, dust. Protruding from each eyelid there are approximately 100 eyelashes.

> Light passes through the cornea and is focused on the fovea by the lens.

c. Eyebrows

The frontal bone protrudes above the orbit of each eye and is covered by a thickened ridge of skin. This in turn is covered with a strip of flattened hairs known as the eyebrows.

d. Conjunctiva

It is an extremely thin transparent mucous membrane which covers the antherior surface of the eye

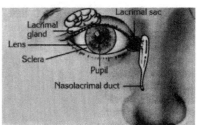

Figure: Tear apparatus.

SENSES

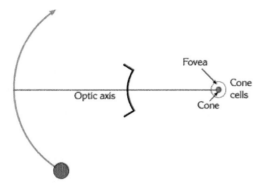

Figure: The location of rods and cones in the human eye can be demonstrated in the following way. The subject is told to look straight ahead. A red ball is brought slowly into view from right to left. The individual perceives first the shape of the ball, then the color. As the ball is swung around to the right, the color detail is lost before the shape. The results of this simple experiment prove that the rod cells are located at the periphery of the retina and the cones are centrally placed.

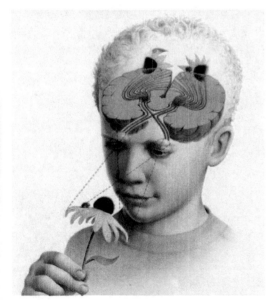

Figure: Light is interpreted in the brain.

e. Tear apparatus:

It includes the lacrimal gland, lacrimal sac, lacrimal duct and nasolacrimal duct. It moistens the eye sphere.

The lacrimal gland, which secretes tears, is located in the upper lateral eyelid and develops in infants during the first months of life. Over this period, the eyes of the newborn should be protected from dust, intense light, and other irritants.

Tears are composed of salt, water, and organic compounds produced by the mucous membrane.

f. Motor Apparatus

A vertebrate has the ability to rotate its eyeball in order to view its environment. This is achieved by six sets of muscles. The movement of the head complements the movement of these muscles. Each muscle is responsible for movement in a particular direction in order to view a single 3-dimensional image. **Any disturbance of these muscles results in double vision**. The movement of these muscles is regulated by cranial nerves

3. Visual Process

The process of perception, transmission and interpretation of vision is performed in five main steps.

Light rays are refracted when they enter the eyes. The image is focused on the retina by the adjustment of the lens and convergence of the image.

Light waves are converted into nerve impulses by neurochemical activity.

Neural activity is processed in the retina and the coded impulses are transmitted through the optic nerve.

The impulses are processed at the brain, the inverted image is corrected and the information is interpreted.

The image formed on the retina is inverted, and it is thought that perhaps this image is righted in the brain by experience. The eyes and the brain working together allow us to see the objects right side up and in three dimensions.

a. Refraction and Accommodation

In order for clear vision to be possible, some systems of focus adjustment for distant and near objects must exist. The ability of the eye to adjust its focus is known as **accommodation**. This is achieved in vertebrates through two different methods.

Accommodation using an elastic lens: The lens of the eye is elastic and bulges or flattens according to the distance of the object focused on (human lenses are elastic).

Accomodation using a rigid lens: A rigid lens cannot change its shape, instead, its distance from the retina is adjusted according to the distance of the object.

4. Eye Defects

The pupil dilates or contracts depending on the amount of light present. Under dark conditions, the pupil dilates to receive the proper amount of light for normal vision. It constricts if the intensity of light is strong.

The activity of the pupil is regulated by reflex actions. At rest, parallel light rays should focus on the retina. There are, however, a number of defects that may affect the eye and cause the image to not be properly focused on the retina. Some of these problems are described below.

a. Myopia (nearsightedness)

While at rest, instead of focusing on the retina, the light rays focus in front of it. This type of eye defect is called **myopia**.

The major cause of this defect is the difference in diameter between the anterior and posterior portion of the eye. In such cases, the posterior portion is wider than the anterior. The eye ball is longer than normal.

This condition can be corrected by wearing glasses or contact lenses with concave lenses. This defect also can be corrected by the latest techniques in laser surgery.

b. Hypermetropia (farsightedness)

At rest, the light rays focus behind, instead of on the retina. This type of eye defect is termed **hypermetropia**. The major cause of this defect is again the difference in the diameter between the anterior and posterior portion of the eye, the anterior portion being wider than the posterior. The eyeball is shorter than normal. The condition can be corrected by wearing glasses or contact lenses with convex lenses.

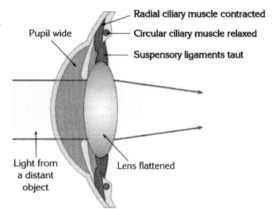

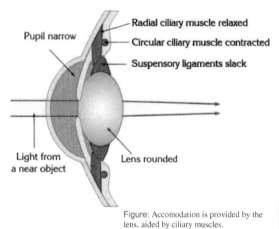

Figure: Accomodation is provided by the lens, aided by ciliary muscles.

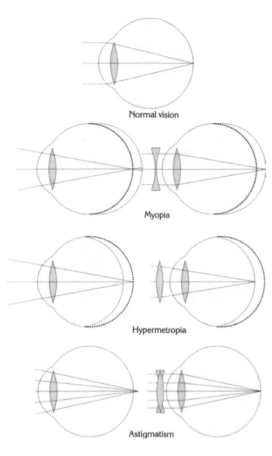

Normal vision

Myopia

Hypermetropia

Astigmatism

Figure: Common eye defects and their treatments.

c. Astigmatism

This describes the condition where the image is constantly unclear due to nonuniformity of the cornea (uneven cornea). This defect can be corrected by wearing cylindrical edged glasses (uneven lenses).

d Prestism

This describes the condition of the lens losing its elasticity due to age. After middle age, the ability of the eye to focus clearly on near objects is reduced. A young man for example, can clearly see an object 10-15 cm in front of him. An old man can only clearly see an object that is more than 80 cm away from his eyes. This problem can be corrected by wearing concave glasses.

e. Color blindness

Complete color blindness is extremely rare. In most instances, a particular type of cone (blue, green, or red absorbing cones) is lacking or deficient in number. The lack of red or green cones is the most common, affecting about 5-8 % of human males. Color blind people are unable to distinguish red color from green. This illness is heritable.

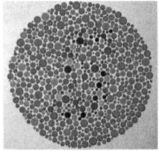 

 In order to have clear vision, light must reach the fovea of the retina. Any problems in the light pathway or eye structure can prevent this. However, many of these problems can be solved and the patient may regain normal vision ability. One of the latest and most effective methods is laser surgery.

THE EARS

The ears, responsible for both hearing and maintaining equilibrium of balance, are composed of the outer ear, middle ear and inner ear. They are extremely sensitive to variations in sound and gravity. Furthermore, they are closely connected to the brain. The ear converts signals carried by sound waves into nerve impulses that it sends to the brain. The ear detect frequencies from 20 Hz (a bee buzzing) to about 18,000 Hz (a very high-pitched whistle)

1. The Structure of the Ears

a. Outer ear

This structure is peculiar to terrestrial organisms and is involved in the collection of sounds and the amplification and transmission of them to the middle ear. The outer ear is composed of:

> **!** Two important sensory functions are accomplished by the ears: hearing and balance.

pinna   external auditory canal   eardrum.

	OUTER EAR	MIDDLE EAR	INNER EAR	
			Cochlea	Sacs and Semicircular Canals
Function	Directs sound waves to tympanic membrane	Picks up and amplifies sound waves	Hearing	Maintains equilibrium
Anatomy	Pinna, auditory canal	Tympanic membrane; ossicles	Organ of Corti; auditory nerve	Saccule and utricle; semicircular canals
Medium	Air	Air (auditory tube)	Fluid	Fluid

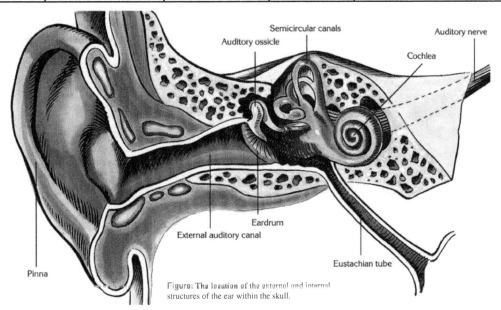

Figure: The location of the external and internal structures of the ear within the skull.

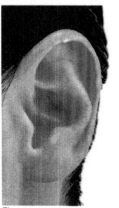

Figure: The outermost part of the ear is the pinna.

1. Pinna

It is a cartilaginous tissue with a distinct shape. The pinna is long and moveable in some mammals. It is however, short and immoveable in human beings. It collects sound waves and determines their source.

2. External auditory canal

It is an "S" shaped canal 3-4 cm in length in adults. Generally, it is formed from both cartilage and bone. The skin of the external auditory canal contains glands that produce a wax-like substance. Furthermore, the entrance of the canal is lined with hairs which filter dust and solid particles.

3. The eardrum

It separates the outer ear from the middle ear. It is a thin, half-transparent membrane, approximately 1 cm in diameter, and composed of collagen fibers. The eardrum has two states: relaxed and contracted. The hammer, anvil and stirrup are the bones located in the middle ear. The bones are connected to each other by moveable joints between the eardrum and the inner ear. The hammer is attached to the eardrum. The anvil acts as a bridge between the hammer and the stirrup, which is attached to the oval window (entrance of the inner ear). The contracted state of the ear drum prevents sound of short wavelength from interfering with sounds of longer wavelength.

b. Middle ear

It is composed of the bones involved in hearing and their associated structures. The Eustachian tube opens into the middle ear. This tube connects the middle ear to the pharynx: it equalizes air pressure on both sides of the eardrum when the outside air pressure is not the same.

1. Ear ossicles

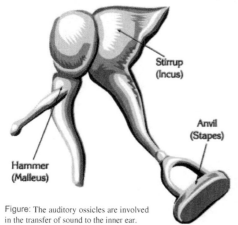

Stirrup
(Incus)

Anvil
(Stapes)

Hammer
(Malleus)

Figure: The auditory ossicles are involved in the transfer of sound to the inner ear.

Three small bones, known as the hammer, anvil and stirrup, are located in the air-filled middle ear. The ear ossicles are attached to each other by moveable joints. The eardrum and ossicles carry the vibrations to the inner ear. The surface area of the eardrum is 25 times that of the surface area of the stirrup. Thus, the stirrup increases vibrations 15-20 times while transmitting them to the oval window. In this way, light vibrations can be detected by the ears due to the increased intensity of sound provided by the ossicles.

2. Eustachian tube

It is 4 cm long and 2 mm wide and forms a connection between the pharynx and the middle ear. The tip of the Eustachian tube of the pharynx is closed by a valve to facilitate better hearing. The valve opens if the air pressure between the external environment and the middle ear is unbalanced. Thus, external pressure is balanced by the Eustachian tube in order to prevent rupture of the eardrum. Yawning and swallowing results in the equilibrium of air pressure in the middle ear.

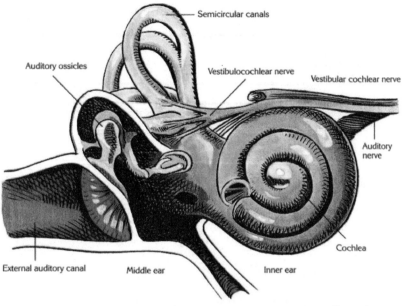

Semicircular canals

Auditory ossicles

Vestibulocochlear nerve

Vestibular cochlear nerve

Auditory nerve

Cochlea

External auditory canal

Middle ear

Inner ear

c. Inner ear:

It is a well-protected organ of the body. Although the outer and middle ears are involved only in hearing, the inner ear is also involved in the maintenance of balance. The inner ear, anatomically speaking, has three areas: the semicircular canals and the vestibule are related to equilibrium; the cochlea is concerned with hearing. There are three types of semicircular canals: anterior, posterior and lateral. These canals have an ampulla at their vestibule end.

Cochlea: It is an osseous (bony) structure in the inner ear involved in hearing. The vestibule, or chamber, lies between the semicircular canals and the cochlea. It contains two small membranous sacs called the utricle and the saccule. Both of these sacs contain little hair cells, whose cilia are embedded within a gelatinous material.

Calcium carbonate ($CaCO_3$) granules, or **otoliths**, rest on this material.

The cochlea resembles the shell of a snail. Three canals are located within the tubular cochlea: the vestibular canal, the cochlear canal, and the tympanic canal. Along the length of the basilar membrane, which forms the lower wall of the cochlear canal, are little hair cells whose cilia are embedded within a gelatinous material, called the tectorial mebrane. The hair cells of the cochlear canal, called the organ of Corti, synapse with nerve fibers of the cochlear (auditory) nerve. The cochlear nerve generates nerve impulses that go to the brain stem and finally to the temporal lobe of the cerebrum, where they are interpreted as sound.

There are approximately 20-40 thousand hairs on the free ends of the organ of Corti. The hairs are positioned into five rows and generate impulses in sensory neurons. Information concerning the pitch of sound is transmitted to the CNS.

Figure: The structure of the components of the middle and inner ear.

 The semicircular canals and vestibule are the parts of the ear responsible for equilibrium. The organ of Corti is sensitive to soundwaves and provides hearing.

 The vestibule is located between the semicircular canals and the cochlea. It has two structures called the utricle and saccule.

The vestibule aids body to be in balance.

SENSES

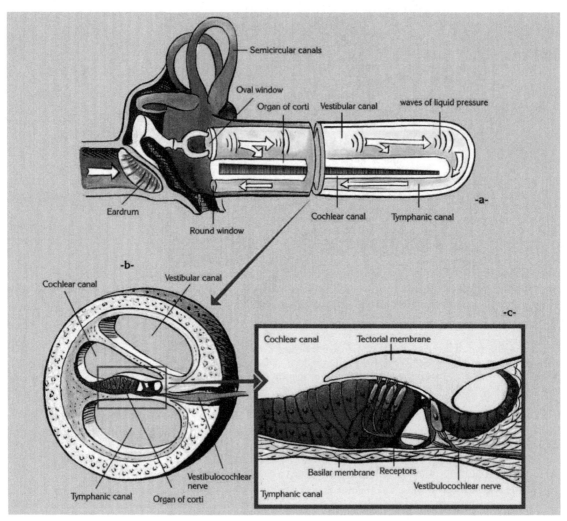

Figure:
a. The structure of the inner ear shown simplified in extended, longitudinal section
b. Transverse section through the inner ear, shown extended and simplified. c. The organ of Corti and its component parts.

The organ of Corti is attached to a fibrous membrane. The fibers of this membrane are under tension in much the same way as the strings of a guitar. The lengths of the fibers vary, and it is accepted that they are adjusted to different sounds. High frequency sounds vibrate short fibers, whereas low frequency sounds vibrate long fibers. Thus, differences in pitch are detectable.

Albino cats are born deaf, as their organ of Corti is undeveloped. The ciliated cells of the organ of Corti synapse with axons of neurons. The acoustic axons originating from the cochlea are regularly ordered according to the frequency they detect, and this same order is maintained in the central nervous system.

2. Hearing

Sound enters the ear through the air. The ability of the air to transfer sound is rapid as compared to its transference through water and solids. Sounds are differentiated physically according to their intensity, frequency and wavelength. The intensity of sound is detected by the ears.

The number of sound vibrations per second is termed the frequency, which determines the tone of a particular sound. The wavelength, however, determines the type of sound. For example, the difference in sound between a piano and violin can be differentiated, even though they have the same frequency and intensity. This can be explained by the fact that the wavelength of each sound is different.

a. The mechanism of hearing:

Sounds are collected into the external auditory canal by the pinna. They travel to the eardrum and cause it to vibrate. Each sound has a particular vibration. Sound is generated by the movement of a solid structure such as wood, glass or metal, and is reflected by solid objects. The eardrum has no level of vibration peculiar to it, which explains why waves are transmitted across the eardrum without any distortion. The vibrations of the eardrum are transmitted to the ear ossicles. They then pass through the oval window. The membranes of the oval window vibrate and transmit this vibration or message to the fluid within the cochlea (perilymph and endolymph). The vibrations of the perilymph influence the membranes of the cochlear duct and transmit the messages to the endolymph.

Finally, the vibrations pass from the endolymph to the organ of Corti, which produces nerve impulses at the cochlear nerve endings. The nerve impulses are transmitted to the brain by the auditory nerve, also known as the acoustic nerve.

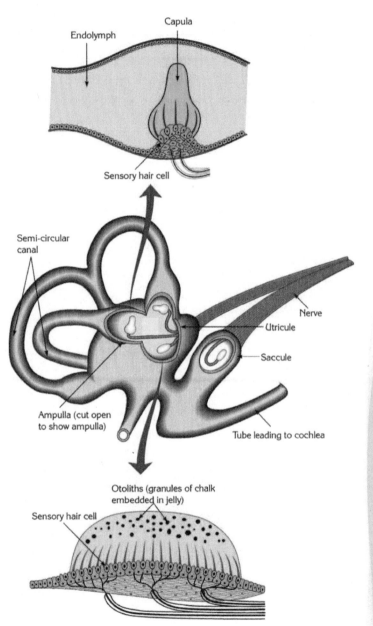

Structurally, the human sound system is a bit of a lash-up--it looks like an improvised temporary solution sketched on the back of an envelope by a genius inventor. A funnel, a tunnel, a membrane, three bones, a window, some fluid, a spiral tube with more membrane, rows of hairs and, finally, pathways to the brain. But another way of looking at it is as a marvel of miniaturization. Compressed into two cubic centimeters are a wide-range sound-wave analyzer, a noise-reduction system, two-way communications, a relay unit, multi-channel transducer and a sophisticated hydraulic balance system.

Noise-reducing bones

The external ear, the bit you can see, is called the pinna, and its job is to direct sound waves down the auditory channel to the eardrum. Sound waves are vibrations of air molecules, and they have amplitude and frequency. Amplitude determines volume: the louder the sound, the higher the wave, the greater the amplitude. Frequency is pitch: the more waves per second, the higher pitched the sound.

We can hear frequencies from 15 to 10,000 cycles per second, up to 20,000 if we're young (Bats and some rodents can detect 115,000 cycles per second.). Our system works best for the middle-range frequencies involved in speech. A sound wave takes 20 milliseconds to pass through the ear. At the end of the auditory canal, the eardrum, or tympanic membrane, vibrates as the sound wave hits it. Your eardrum is incredibly sensitive in response to very faint sounds, it moves as little as one hundred-millionth of a meter. Air pressure on either side of the eardrum must be equal. That's why our ears pop in aiplanes, as the Eustchian tube, leading to the back of the throat, opens to equalize the pressure.

Next stop for the sound wave is the middle ear, which contains the ossicles, the three smallest bones in the body. The malleus (or hammer) picks up the vibrations from the eardrum and bangs on the incus (anvil), which is connected to the stapes (stirrup). This curious mechanical set-up can work as a noise reduction system. About a tenth of a second after hearing a loud noise, muscles attached to the three bones contract, damping down the sound. The same thing happens before you speak. If it didn't, a crying baby would deafen itself. This system also reduces lower frequencies, which would otherwise swamp the higher notes found in speech.

The vibrating stirrup is attached to the membrane of the fenestra ovalis (oval window), which it rattles, passing on the reduced vibration to the inner ear. This is the point at which the sound moves from the air to the water: into the liquid-filled channels of the cochlea. A pea-size, spiral-shaped bony fortress buried in the thickest part of the skull, the cochlea is a mechanism of mind-boggling complexity.

At its heart are hair cells, special hearing receptors that are among the most remarkable structures in the body. These hairs turn vibrations into electrical impulses to send to the brain. There are 32,000 sensory hair cells in each ear--not a lot compared with an eye's 300 million light receptors--and you start losing them from birth, a process speeded up by loud noise, drugs and illness.

Two membranes, the basilar and the tectorial, form a partitioned-off triangular canal running inside the cochlea down the length of its spiral. The beating of the stirrup sets the basilar membrane vibrating, which distors the sensory hair cells.

The system is extraordinarily sensitive, detecting sounds that move hairs by a three-thousandth of a degree. It's also remarkably fast. Our brain can distinguish when a sound reaches one ear six to ten millionths of a second after the other. Amazing, considering that elsewhere in the body, chemical transmission crawls along at mere thousandths of a second. The answer may lie with a newly discovered set of gates called "tip links", which open up the ion channels between nerves instantaneously.

Another of the mysteries of the cochlea is how it handles very high-frequency sounds--20,000 cycles per second and beyond. The basilar membrane sorts out the great tangle of sound into frequency bands: high frequencies at the base of the spiral shell, low frequencies at the apex, at intervals of about a third of an octave every millimeter.

Feedback loop?

The membrane vibrates at frequencies a hundred times faster than any movement in the fluid. Recent discoveries about auditory nerve pathways suggest that there may be a feedback mechanism between the brain and cochlea making the membrane move faster. It has also been found that some sensory hair cells move independently.

Once nerve impulses begin to travel from the cochlea to the brain the mysteries are far from over. Very impressive, for instance, is the way the ear is able to discriminate so sensitively, despite the fact that the auditory nerve contains about 28,000 neurons, compared with the optic nerve's one million.

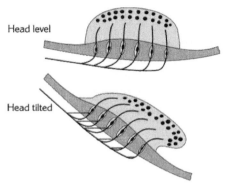

Head level

Head tilted

Figure: How a macula works: When the head is level the otoliths press downwards. When the head tilts, speeds up, or slows down, the otoliths pull against the hair cells which send impulses to the brain.

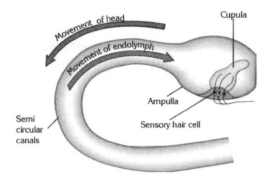

Movement of head

Movement of endolymph

Cupula

Ampulla

Semi circular canals

Sensory hair cell

Figure: How semi-circular canals work: When the head moves one way, endolymph is 'left behind' and apparently flows the other way, bending the cupula. This stimulates the hair cells, which send impulses to the brain.

b. Determination of Sound Direction

The direction of sound can be determinated if it enters the ear close to its point of origin. The pinna of animals plays an important role in the determination of the origin of sound and responds by rotating in its direction. Humans however, move their head towards the origin of the sound.

c. Vestibular Organ and Semicircular Canals

The vestibule of the inner ear (the part between the cochlea and semicircular canals) detects our position with respect to gravity (static equilibrium). There are three semicircular canals, each of which is oriented in a different plane in space. Movement in any plane can be detected by at least one of the canals. The brain interprets complex movements by comparing information sent from each canal. The semicircular canals detect the direction of our movement as a result (dynamic equilibrium).

Figure: Sense of hearing has an essential importance in communication.

READ ME

BALANCE

Investigation into the control of balance by a frog;

As shown in the figure, a frog is placed on a piece of wood and covered by a beaker. First the frog is kept in a vertical position. The wood is then moved slightly downward and then upward. During these movements the frog adjusts the position of its head depending on the position of the wood.

The movements of the frog to maintain its balance are reflexes performed by the contraction and relaxation of the head and neck muscles. Sense organs which control these reflexes are situated within the inner ear.

In the event of any change of position, sensory organs are stimulated and the nervous system transports the stimulus to the muscles of the head and neck and other related parts of the body.

If the inner ear is damaged, these movements are not observed.

In the control of balance, other sense organs may play a role, such as receiving stimuli from the muscles and eyes.

SKIN

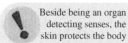

Beside being an organ detecting senses, the skin protects the body as a mechanical barier, excretes some metabolic wastes, and regulates body temperature.

All multicellular organisms have a skin composed of one or more layers. The skin functions as a protective layer for these organisms. The components of skin vary according to the type of organism, and may contain hairs, nails, scales, sweat glands or sensory receptors.

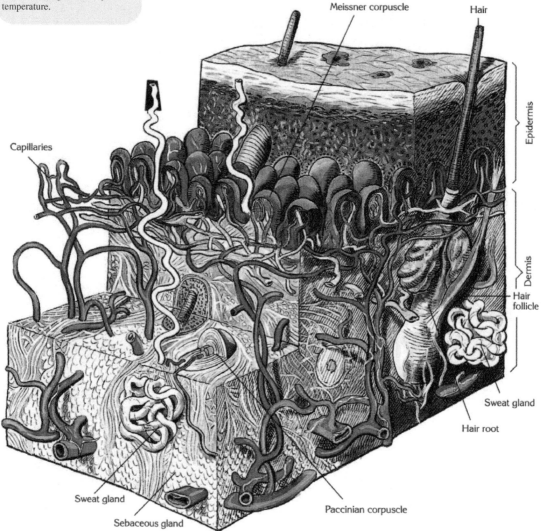

Figure: A 3-dimensional cut-away section of skin and its components.

The functions of skin can be listed as follows:

protection of the inner layers of the body from physical and chemical effects

prevention of the entry of microbes into the body prevention of water loss in terrestrial organisms

protection of cells from the adverse effect of ultraviolet light

regulation of body temperature in higher organisms. In hot conditions, capillaries in the skin dilate and radiate heat. The same capillaries constrict in cold conditions to prevent heat loss

a site of gas exchange

excretion of catabolic wastes via sweat glands

maintenance of a moist body surface secretion

of fat

absorption of some medicines

Skin is composed of two completely different layers:

Epidermis

Dermis

The epidermis is generated from ectoderm and is composed of epithelial cells. The dermis is generated from mesoderm and is located directly beneath the epidermis.

1. Epidermis

The outermost layer of the body is known as the epidermis. It is composed of keratinized epithelial cells and its surface is covered by either a layer of secretions from epidermal glands or by keratinized cells. This layer is involved in the protection of the body from physical and chemical hazards and forms a barrier against microbes.

The epidermis contains no blood vessels, thus it is supplied with nutrients diffusing from blood vessels in the dermis. The epidermis is thickest in the palm of the hand and the sole of the foot.

The epidermis is approximately 0.7 mm in thickness. The upper section of the epidermis is composed of nonliving epithelial cells containing a polypeptide known as keratin.

This layer is shed over a period of time and is replaced by the rapidly dividing cells of the inner layer. The color of the skin is conferred by melanin, produced by granules located in melanocytes of the epidermis.

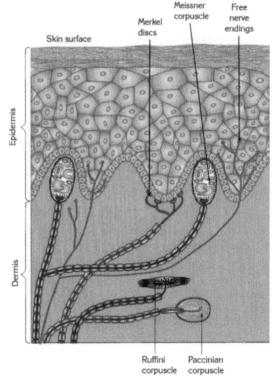

Figure: Special sensory receptors found in skin.

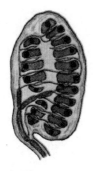

100 μ Meissner corpuscle   4mm Paccininan corpuscle

50 μ Krouse corpuscle   200 μ Ruffini corpuscle

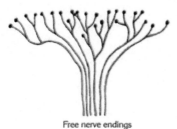

Free nerve endings

Figure: Sense organs of the skin and their size.

Within the urine, some water, minerals and urea are removed from the body. Each day more than half a liter of sweat is excreted.

2. Dermis

It is a layer located directly beneath the epidermis and is 4 mm in thickness. It is composed of interconnected collagen and elastic fibers of connective tissue. The dermis is rich in blood vessels and nerve endings. The receptors located in the skin are connected to these nerve endings.

Blood vessels are present in this layer and are involved in supplying nutrients to cells and in thermoregulation, the regulation of the temperature of the skin. Furthermore, the dermis contains smooth muscles, sweat and sebaceous (oil) glands, hair follicles, touch receptors and lymph vessels.

Smooth muscle is connected to each hair follicle and is involved in hair movement. The skin is also connected to sympathetic nerves (that is why our hairs erect under panic situations). Touch receptors are located in the upper surface of the dermis and the lower surface of the epidermis, sometimes at the junction between them. Touch receptors, either Pacinian corpuscles or Meissner corpuscles are involved in sense reception. Pacinian corpuscles are sensitive to high pressure, while Meissner corpuscles are sensitive to light pressure. Other receptors in the dermis are sensitive to temperature.

3. Accesory Structures of the Skin

a. Skin glands

The skin includes two types of glands: sebaceous glands and sweat glands.

Sebaceous glands: The sebaceous glands are scattered through the skin. They are present in all areas of the body except the palms and soles. They secrete sebum, a mixture of fatty material and cell debris, either into hair follicles through a short duct, or directly to the surface of the skin. Furthermore, sebum functions as a barrier against infectious bacteria and fungi. There are between 400 and 900 sweat glands per cm^2 on the face and head.

Sweat glands: They are present in nearly all regions of the skin and consist of tiny tube-like structures. The sweat glands of the dermis are coiled and ball-shaped.

They open onto the surface of the skin via the pores. They are involved in the removal of water, minerals, urea and other substances by sweating. The composition of sweat resembles dilute urine. The sweat glands are therefore considered a third kidney. Approximately 500-600 ml of liquid is sweated in a 24 hour period. A heavy laborer working in high temperatures may lose 15 litres of moisture per day. Furthermore, odors are secreted from apocrine glands, a type of sweat gland. They are concentrated in the armpits, the groin and around the nipple region.

The main function of sweat glands is the regulation of body temperature by evaporation of water. The body temperature remains constant since excess heat is released by evaporation of water through sweating. If the skin temperature were unable to be regulated, for example in the event of third degree burns, this would prove fatal. Dogs have no sweat glands in their skin, only on their tongue. This fact explains why dogs protrude their tongue when they are overheated.

b. Hair follicles

This feature is unique to mammals. Hair follicles cover the whole body except for the palms, soles, lips, etc.

The root of the hair in the dermis is termed the hair follicle, whereas the visible portion of it is termed the hair shaft. The diameter of the hair in the follicle is wider than that of the shaft. Hair is formed from keratinized epithelial cells. In this process, living hair cells are impregnated with keratin from epithelial cells. Each hair follicle has a sebaceous gland in addition to smooth muscle. During contraction of the smooth muscle, the hair is raised from its normal position at the surface of the skin to an erect position.

Hair color is determined by pigment produced by melanocytes located at the base of the hair follicle. The greater the amount of melanin produced, the darker the hair color. Trichosiderin is an iron pigment that gives red color to hair.

Blonde hair color results from less melanin production, the amount of production being genetically controlled. If the gene that controls melanin synthesis is nonfunctional, the individual becomes albino, with colorless hair and skin. The hair loses its color due to aging or stress as the production of melanin decreases.

The distribution, quantity, length, diameter and appearance of hair varies in different regions of the body.

c. Nails

They form protective coverings of the toes and fingers. Each nail is composed of a nail plate and a nail bed. The nail plate is a continuation of the epithelium of the skin, and is therefore composed of epithelial cells.

The base of the plate resembles the shape of a half-moon and contains rapidly dividing epithelial cells. After division, impregnation with keratin forms a horny structure. Growth of the base pushes the nail plate forward. During normal use, the extended portion is worn away.

The thickness of a nail is between 0.5-0.7 mm. Nails grow between 0.5-1 mm per week. The amount is affected, however, by hormones, disease and diet.

d. Skin Pigments

Skin color is produced by granular pigments. Body color has different functions in vertebrates and invertebrates. Some, for example, use color in adaptation to the environment. Color can be used in defense and in the attraction of a mate. The main function, however, is protection from harmful rays of sunlight.

The root of the hair is called the **hair follicle**, and its visible part is the **hair shaft**.

In emergency or panic situations, smooth muscles found in the dermis contract. As a rusult, hair is raised from its normal position and becomes erect.

Both hair and nails are formed from keratin-containing epithelial cells.

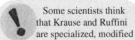
It is interesting that some insects mimic the color and design of their environment when they are forming their new exoskeleton.

The production of color results from the distribution or accumulation of pigments in the body. The body darkens in color when the pigments are accumulated at the core of the cell. The pigments located in the star-like cells of amphibia are known as chromatophores, and are rare in mammals.

Only albinos lack melanin in their bodies. The quantity of melanin in various races of humans, the quantity increasing from the poles to the equator, is an adaptation to protect the body from intense sunlight. Pigmentation is regulated by both the endocrine and nervous system. Sunlight entering the eye stimulates the secretion of melanocyte stimulating hormone (MSH) from the pituitary and is involved in the accumulation or distribution of pigments.

4. Touch Receptors

The receptors of the skin are involved in the perception of stimuli from touch, pain, temperature, pressure and vibration.

These receptors are categorized into two groups: Paccinian corpuscles and Meissner corpuscles.

Paccinian corpuscles are involved in the reception of heavy pressure, whereas Meissner corpuscles are involved in the reception of light pressure. Pain is detected only by nerve endings.

a. Noncapsulated receptors

Free nerve endings, Merkel's corpuscles and hair follicle receptors are noncapsulated receptors of the skin. Free nerve endings are involved in pain, light, touch, pressure and probably temperature sensation.

Merkel's corpuscles are involved in reception of pressure and light touch. They are present in deep epidermal layers of the palms and soles.

Hair follicle receptors are involved in the reception of touch in the region of the nerve net around the hair follicle.

b. Capsulated receptors

Meissner corpuscles, Paccinian corpuscles, Krause corpuscles and Ruffini corpuscles are the capsulated receptors of the skin.

Meissner corpuscles are involved in reception of touch in the palm, sole and lips.

Paccinian corpuscles are involved in reception of mechanical stimuli. They are pressure receptors located deep in the dermis

Krause corpuscles are involved in reception of cold and pressure.

Ruffini corpuscles are involved in reception of heat, touch and pressure.

SMELL

The nose is the organ of the body involved in both respiration and smell. The reception of smell takes place in chemoreceptors located in the **nasal cavity**. The total surface area of chemoreceptors in the nasal cavity is 10 cm^2. This region is known as the **olfactory region** (oleo: smell; facio: making). The human nasal mucosa may contain 25 million receptor cells, whereas that of a dog contains 220 million.

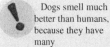

Dogs smell much better than humans, because they have many more chemocereptors in their nasal cavity to detect smell.

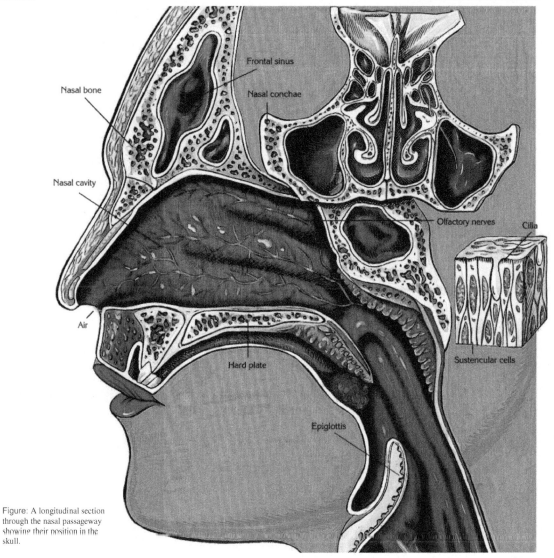

Frontal sinus

Nasal bone

Nasal conchae

Nasal cavity

Olfactory nerves

Cilia

Air

Sustencular cells

Hard plate

Epiglottis

Figure: A longitudinal section through the nasal passageway showing their position in the skull.

57

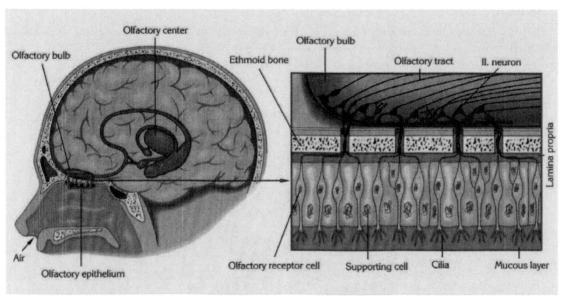

Figure: The structures involved in smell.

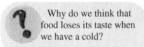

Why do we think that food loses its taste when we have a cold?

When you have a cold you think that food loses its taste, but actually you lose the ability to sense its smell. When you smell something, some of the molecules move from the nose down into the mouth region and stimulate the taste buds. Therefore, part of what we refer to as smell may actually be taste.

The olfactory receptors, like those of touch and temperature, adapt to outside stimuli. In other words, after a while, the presence of a particular chemical no longer causes the olfactory cells to generate nerve impulses, and we are no longer aware of a particular smell.

The sense of taste and smell suplement each other, creating a combined effect on the cerebrum. For example, when you have a cold, you think that food loses its taste, but actually you lose the ability to sense its smell. When you smell something, some of the molecules move from the nose down into the mouth region and stimulate the taste buds there. Therefore, part of what we refer to as smell may actually be taste. Olfactory receptors detect different smells because of the specific binding of airborne gases with receptor chemicals located within the cilia.

By means of olfactory nerve, the impulse travels to the olfactory area of the cerebral cortex to be interpreted. On its way, it tavels through the brain's limbic system, the area of the brain responsible for many drives and emotions. Does the odor of a piece of cake cause you pleasure? That is your limbic system at work.

1. Mechanism of Smelling

The ends of the olfactory nerves within the nasal cavity are ciliated and are exposed to the air. Any odors in the air are detected by these projections. The particles in inhaled air are dissolved in the fluid (mucosa) that covers the receptors in the nasal cavity, and olfactory nerve cells are stimulated by these dissolved chemicals. Smelling is fundamental in the detection of food, maintenance of relationship, reproduction and communication of some animals.

The ciliated structures in the nasal cavity converge to form the olfactory nerves, which collectively form the first cranial nerve. Messages are transmitted to the olfactory center located in the temporal lobe of the cerebrum for interpretation.

The ability to detect smell is more advanced in terrestrial organisms as compared to aquatic organisms. The level of detection in humans however, is less advanced when compared with other organisms. The smelling ability of dogs, for instance, is 10 million times that of humans.

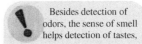 Besides detection of odors, the sense of smell helps detection of tastes, maintenance of relationship, reproduction and communication of some animals.

READ ME

SMELL

Of all the senses, smell provides our most direct contact with the environment. Every time you inhale, you bring microscopic pieces of the outside world into physical contact with the nerves in your nose for chemical analysis. These nerves are unique in the body in that they have one end dangling in the outside world and the other feeding into the brain.

Together with taste, smell is our oldest sense. All living creatures, from single-celled organisms to human beings, can sense chemicals. The origins of smell lie several billion years in the past, when the first bacteria learned to gravitate towards food and away from toxins. It is estimated that as many as 1000 of our 100,000 genes are occupied exclusively with producing smell receptors, compared with the three genes that control color vision.

From a survial sense...

As primates and then humans evolved, smell was crucial to survival for finding food and mates, and escaping poisons and predators. Yet in modern times, this sense has lost a good deal of its importance. About four million Americans suffer from complete loss of smell--anosmia--which may be genetic, caused by a blow to the head, or the result of an infection, drugs or smoking.

Smell is about 10,000 times more sensitive than taste. With practice, humans can detect 10,000 different smells, but we're insensitive compared with the rest of the animal kingdom. For a human to smell vinegar, there must be 500,000 million molecules of acetic acid per cubic meter of air, while a dog needs only 200,000. Of course, some molecules are smellier than others. We can pick up a single molecule of mercaptan--the active ingredient in skunk smell--in 30,000 million molecules of air.

For most of us, smell is an occasional sense. You don't notice it, then suddenly it hits you--new-mown grass, the aroma of coffee, a perfume, some fish that has gone off--and for that moment it dominates your awareness. There is an immediacy about smell that's missing in the other senses, and everyone has experienced the way a smell can vividly spark off memories. These features--immediacy and permanence--are rooted in how our sense of smell works.

Compared with the Heath-Robinson complexity of the other senses, the detection of smell is simplicity itself. Deep inside each nostril is a 2.5-centimeter-square patch of five million "olfactory cilia". These threads hang in clusters of eight or so at the end of olfactory rods, which lead to axons running directly into the brain.

Cilia don't wave about like the hairs lower down, but are submerged in mucous. Unlike other brain cells, which last a lifetime or die, olfactory nerves are renewed every one or two months. So how does a neuron translate the presence of a molecule outside the cell into an electrial signal inside the cell? Research points to G-proteins, found at the base of the smell receptors, which act like an amplifier, turning the small stimulus of an odor molecule into a major cellular reaction.

Why smells stir deep emotions becomes clear when you see the route a smell signal takes to the brain. First stop are the olfactory bulbs at the base of the brain. From there the signal moves to the limbic system - an ancient part of the brain concerned with hunger, moods, sexual urges and emotions such as fear - and then to the hippocampus, seat of memory.

The signal then spreads to the frontal lobes, which give rise to conscious thought. Until a few years ago smells were thought to belong to various categories - ethereal, aromatic or ambrosial - each of which researchers believed had its own receptors.

But the latest thinking is that each smell causes different cells in the olfactory bulb to fire in a unique pattern. Research on locusts, whose smell systems are similar to our own, shows that different smells produce synchronized oscillations in different parts of the bulb, and a single neuron can respond differently to several smell patterns.

That is certainly how the latest "electronic nose" works. The Aromascan creates a graphical computer "smellprint" by using the fact that the electrical resistance of a chemical compound varies depending on the type of gas molecules in contact with it.

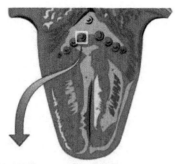

Filamentus papilla   Round papilla

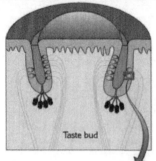
Taste bud

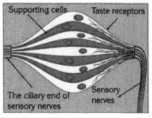
Supporting cells   Taste receptors

The ciliary end of   Sensory
sensory nerves   nerves

Figure: The tongue contains two different types of papillae. Filamentous papillae form a surface to which moistened food particles become attached. Round papillae form the taste buds. In longitudinal section, the individual taste receptors are visible as seen in the diagram.

TASTE

The tongue is one of the most important organs of speech and nutrition in human beings. It contains approximately 9-10 thousand taste buds which are used to differentiate the tastes of foods.

The tongue of a child contains more taste buds than that of an adult. The taste buds are oval-shaped, 50-80 microns in length, and are embedded in the epithelial layer of the tongue. In the structure of a single taste bud, there are 5-18 taste hairs supported by cells. The hairs are connected to chemoreceptor cells in the buds.

The tongue is divided into four regions involved in the sensation of taste: bitter, sour, salty and sweet. Sweet foods are tasted at the tip of the tongue, whereas bitter foods are tasted at the rear. Sour and salty foods are simultaneously tasted on both sides of the tongue. Individual tastes can not be differentiated if the nose is blocked and chewing ceases.

All sour foods contain acids, such as citric acid. H^+ ions and acid groups stimulate the chemoreceptors in the regions detecting sour taste.

A salty taste is provided by Cl^- ions in the ingested food.

Common salt is composed of $NaCl^-$ compounds. The chloride (Cl–) -free salts such as sodium sulphate, sodium iodine and bromide, may give a salty taste.

A sweet taste is provided by the chemical content of foods such as carbohydrates.

A bitter taste is provided by many chemicals such as alkaloids and tannins.

Taste is the simplest of the senses. While smell can deal with thousands of combinations, taste--like a good, plain cook--settles for four basic ingredients: sweet, sour (like lemon juice), salt and bitter (like black coffee). Taste receptors on the tongue--the taste buds or papillae--specialize in one or another of these.

The front of the tongue perceives sweet tastes, the back perceives bitter, sour and salt tastes are detected by the sides, at the back and front respectively.

Taste evolved as a warning system, not an aid to gourmet delight, so it's not surprising we are more sensitive to potentially dangerous tastes--usually bitter--rather than sweet ones. In fact, bitter flavors can be picked out at one part in two million, whereas sweet flavors need a concentration of one in 200. Some people have as many as 10,000 taste buds, others as few as

500. But tasting is a tough business: bud cells live only about ten days before replacement.

1. The function of the tongue

Motor function: Smooth muscles confer the shape of the tongue, and these longitudinally and horizontally positioned smooth muscles are involved in the movement of the tongue, generation of specific sounds in speech, chewing and swallowing.

Sensory functions: As explained above, the tongue is involved in the detection of bitter, sour, salty and sweet tastes.

The taste buds for different tastes are found almost everywhere on the tongue. However, certain places include one type of bud more than the others. That is why the back of the tongue can detect bitter taste better than other tastes.

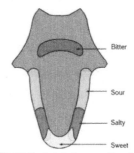

Figure: The taste buds are positioned at the edge of the tongue. Their location is ideal for maximum perception of taste.

READ ME

TOUCH

A sense that's easily fooled

Recent studies show that if you block cold-specific nerve fibers, then touch the skin with something cool, it feels hot. The reception of hardness, texture, dryness and so on is not distrupted.

There are different receptors for different stimuli: for pleasantly warm and for painfully hot, for high and low frequency vibrations, for steady pressure and for light touch--and, of course, for pain.

At two square meters in area, the skin, containing its receptors, is the largest organ of the body. Distributed unevenly across the skin are 200,000 receptors for cold, 500,000 for touch and pressure, and 2,800,000 for pain.

READ ME WISDO

A blind boy does his math homework using Braille--but the real multiplication is going on in his head. Braille is hard to learn at first, but the number of brain cells devoted to the fingertips increases with experience, providing more data-processing capacity.

Filigree of nerves

The human hand is a web of nerves-- muscle-controlling motor nerves as well as the sensory nerves that carry touch signals to the brain. Our fingers must be strong enough to manipulate their environment as well as serving as antennae to explore it.

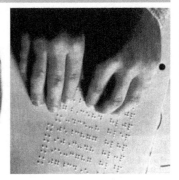

Fingertips have receptors called Pacinian corpuscles, surrounded by fine "onion skin" membranes, which increase the sensitivity.

SELF CHECK

SENSES

A. Key Terms

Lens	Pupil
Choroid	Cones
Fovea	Organ of Corti
Semicircular canals	Taste buds
Chemoreceptors	Vestibule
Pinna	Ossicles

B. Review Questions

1. What are the six types of receptors in the human body?

2.

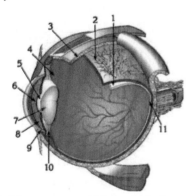

Label the components of the eye in the figure.

3. Explain the functions of the fluid filling the eye chambers.

4. List the accessory organs of the eye.

5. Determine the cause of color blindness.

6. What are the eye defects and their treatments?

7. How is an image accommodated on the retina?

8. Explain the function of the Eustachian canal.

9. List the structures at the inner ear involved in balance.

10.

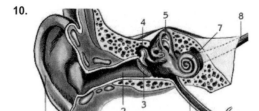

Label the structures of the ear.

C. True or False

1. Taste buds are structures containing the receptors of the sense of taste.

2. Choroid is the pigmented middle layer of eyes that includes blood vessels.

3. Light first passes through the cornea.

4. The part of the retina on which the image is focused is called the blind spot.

5. Calcium carbonate granules found in the utricle and saccule of the vestibule in the inner ear, which help the body maintain balance, are called otoliths.

D. Matching

a. Rod ( ) Small bones of the middle ear.

b. Mechano-receptor ( ) Light receptors which are active in dim light. They detect motion but not color.

c. Proprio-ceptor ( ) The structure found behind the iris which brings light rays to focus on the retina.

d. Lens ( ) Sensory receptor in the muscles that assist the brain to known the position of the limbs.

e. Ossicles ( ) Sensory receptor that is sensitive to stimuli like pressure and sound waves.

Endocrine System

ENDOCRINE SYSTEM

The endocrine system is one of two regulatory systems of the body. It controls body functions by means of chemicals called HORMONES.

As we mentioned before, organisms live in unstable internal and external environments. However, in order to continue to survive, the body requires an internal balance called homeostasis. Homeostasis is established by all parts of the body when they work together in harmony.

The nervous and the endocrine systems connect all systems of the body to each other, make them aware of each other and provide homeostasis. The endocrine system is composed of endocrine glands. The endocrine glands are also known as ductless glands, because they don't have any tubular connections leading to the targets of their secretions. Inside of our bodies we also have exocrine glands. They produce substances which are usually used in digestion. Exocrine glands are directly connected to their targets, in body cavities or even outside of the body, via a tube.

Mammary glands and salivary glands are two well-known examples of exocrine glands. The gonads and the pancreas are endocrine glands that also have exocrine activities. The secretions of the endocrine glands, known as hormones, are transported to their targets by diffusion (if the target is near) or in the blood. They regulate the function of the target organ.

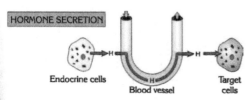

HORMONE SECRETION

Endocrine cells

Blood vessel

Target cells

Figure: Hormones are secreted by endocrine glands and distrubuted throughout the body via the blood.

Generally, hormones have a pronounced effect on metabolic functions; development, production, the level of glucose in the blood and on the concentration of minerals and water at specific levels. They also affect the permeability of the cell plasma membrane.

Hormonal regulation is seen in both plants and animals. Invertebrates, such as segmented worms, insects, molluscs and crustacea, secrete hormones. All vertebrates regulate their metabolic functions by means of hormones.

Hormones are extremely effective chemical substances due to their great potency. For this reason they are found in small amounts in the blood and the urine.

Structurally, many of the hormones are proteins. They can be isolated from the blood and their chemical composition identified. A particular hormone can then be synthesized in vitro and the synthetic product used in the treatment of disorders caused by either a deficiency or an excess of that hormone.

1. Hormones According to Their Chemical Structure

There are two main types of hormones, according to structure.

Exocrine glands have ducts and carry their secretions into body cavities via these tubes. Endocrine glands produce hormones. Hormones are carried to the target organs via the bloodstream.

Peptide hormones are made of amino acids or their derivatives. They can not pass through the cell membrane. They connect to specific receptors on the membrane and trigger an increase of a secondary messenger compound within the cell, such as cAMP. The secondary messenger in turn activates enzymes that alter the cell's function.

Steroid hormones are derivatives of lipids. As a result, they can pass through the cell membrane without the aid of a receptor molecule.

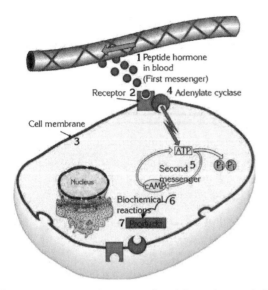

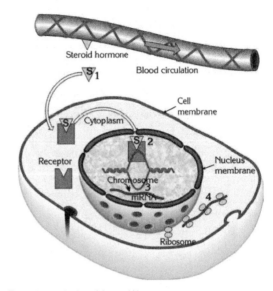

Figure: The effect of peptide hormones on cell metabolism can be summarized as follows:
1. Hormonal or nerve impulse (primary signal)
2. Receptor.
3. Cell Membrane.
4. The reaction of adenylate cyclase with proteins of the cell membrane.
5. Synthesis of cAMP by ATP with adenylate cyclase.
6. Transmission of the hormonal or nervous stimulus to the cytoplasm.
7. Acceleration of biochemical reactions by specific enzymes.
8. Inhibition of production and action of cAMP.

Figure: The mechanism of the steroid hormones:
1. Steroid hormones pass through the membrane and bind directly to receptors in the cytoplasm.
2. The receptor-hormone complex enters the nucleus and initiates the synthesis of mRNA.
3. The assembled mRNA leaves the nucleus.
4. Protein synthesis occurs at the ribosomes.

Inside the cell, they bind with receptor molecules. The hormone-receptor complex then enters the nucleus of the cell, where it acts on DNA to produce proteins. These proteins (enzymes) control certain body processes.

2. Target Organs of Hormones

Hormones must be transported in the blood for maximum effectiveness. Those secreted by endocrine glands usually affect one type of cell or tissue, some may affect a few and some may affect many different types of cells or tissues. For example, ACTH, secreted by the pituitary, is distributed to all tissues of the body in the blood. However, it only affects the cells of the adrenal glands.

Secretin hormone, secreted by the small intestine, affects only the pancreas. Gastrin, secreted by the stomach, affects the gastric glands. The target organs of the pituitary hormones FSH (follicle stimulating hormone) and LH (luteinizing hormone) are the ovaries.

However, the targets of growth hormones are all tissues of the body. Thyroxine in particular has an effect on the metabolism of all cells. Research has shown that there is a specialized receptor protein on the surface of or within the target organ.

 Peptide hormones are detected by receptors located on the cell membrane. Then, a secondary messenger is activated and it alters cell activities. However, steroid hormones can pass through the membranes and have a direct effect on DNA without using secondary messengers.

Once hormone-receptor contact has been established, the effect of the hormone becomes apparent, as explained previously

Each cell gains its own special identity during differentiation. If a particular hormone or receptor cannot be synthesized, or if a secretory gland is destroyed by disease or by accident, many abnormalities can occur.

This is also the case if there is an irregularity in secretion during development. The organism can be affected by many anomalies. In the absence, over-secretion or deficiency of a hormone, irregularities result that may in some cases be fatal.

Hormone and enzyme production in all organisms is internal. Normally, hormones or enzymes are not ingested or injected. If hormones are taken as drugs, the functions of the endocrine system can become disrupted and adjustment of hormones to the system becomes more difficult. For this reason, hormone treatment involves risks to the patient.

3. The Regulation of Hormone Secretion

Hormones are synthesized by endocrine glands when required. Hormone synthesis is usually regulated in the following ways:

An increase or decrease in the amount of different substances (minerals, water, etc.) in the blood.

The effect of one endocrine gland on another, according to the amount of hormone in the blood.

The effect of the nervous system on the endocrine system.

Feedback mechanisms play an important role in the regulation of secretion. This regulation mechanism can be of two types:

A hormone secreted by the endocrine system causes the secretion of a substance by target cells into the blood. This change in the composition of the blood causes a decrease in hormonal secretion from the endocrine gland by affecting its function. For example, parathyroid hormone secreted by the parathyroid gland stimulates the release of calcium into the blood, affecting bone tissue.

The resulting increase in blood calcium levels causes a decrease in secretion from the parathyroid gland. This type of mechanism is known as a negative feedback mechanism. If the amount of calcium decreases in the blood, secretion from the parathyroid gland increases. This is an example of a positive feedback mechanism.

RF (releasing factor), secreted by neurosecretory cells of the hypothalamus, stimulates hormonal secretion by the pituitary. It then stimulates hormone secretion by other endocrine glands.

If the amount of such a hormone increases in the blood, secretions from the hypothalamus and pituitary glands decrease. If the amount of hormone decreases, the secretion of the hypothalamus and pituitary increases.

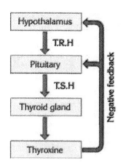

Figure: When thyroxine reaches a certain level in the blood, the hypothalamus is sitimulated and the thyroid stop producing thyroxine.

Hormonal regulation is based on negative feedback mechanisms. That is, if the hormone reaches a certain level in the blood, the hypothalamus and then the pituitary detect this change and trigger the gland to stop its secretion.

THE HUMAN ENDOCRINE SYSTEM

1. Endocrine Glands in the Human Body

The endocrine system in humans is made up of the hypothalamus, pituitary gland, thyroid gland, parathyroid gland, pancreas, ovaries, testes, thymus, epiphysis, adrenal, intestinal and gastric glands. From among all these glands, only the pancreas and the ovaries function as both endocrine and exocrine glands (such glands are called complex or compound glands). The pancreas is a paracrine (the target is far from the gland) gland and gonads are autocrine glands (it is the target of its own secretion. For example male gonads produce testesterone, which triggers sperm production in the seminiferous tubules of the gonads).

a. Hypothalamus

The hypothalamus is a structure in the diencephalon. It plays an important role in the secretory system of the body. It controls the pituitary gland and secretes hormones simultaneously. Its functions include the regulation of body heat and blood pressure, the regulation of sexual desire, the regulation of water in the blood and formation of a regular sleeping pattern, as well as controlling the endocrine system via the pituitary. The secretions of the hypothalamus are known as neuro-hormones.

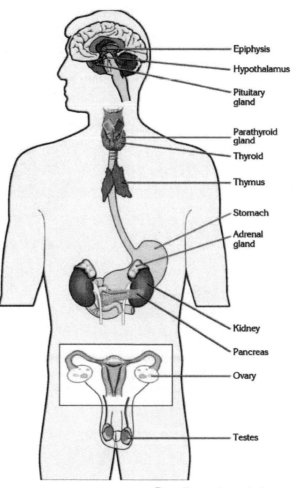

Figure: Human endocrine glands.

HYPOTHALAMUS HORMONES	
Growth hormone releasing hormone	GRH - (GH-RH)
Adrenocorticotropic hormone releasing hormone	CRH - (ACTH-RH)
Thyroid releasing hormone	TRH - (TSH-RH)
Gonadotrophic releasing hormone	GnRH. (LH-FSH-LTH-RH)

The hypothalamus sends hormonal stimulators or chemical substances by way of the blood to the anterior lobe of the pituitary. The secretory mechanism of the pituitary is controlled by the hypothalamus.

Some cells in specific centers of the hypothalamus secrete neuro-hormones, which are also known as releasing hormones, or releasing factors. The relationship between the hypothalamus and pituitary gland is established by these hormones. The hypothalamus has the capacity to release five hormones, and it is believed that inhibitory hormones exist for each pituitary hormone.

b. Pituitary Gland

The pituitary is a gland which is connected directly to the hypothalamus region of the brain via a thin stalk. It is a very powerfull gland and produces many hormones. However, it is amazingly tiny, about the size of a marble or a pea (1 cm in diameter) and weighs between 500-600 mg. The pituitary is an important gland which controls hormone related functions.

The pituitary is composed of three lobes:

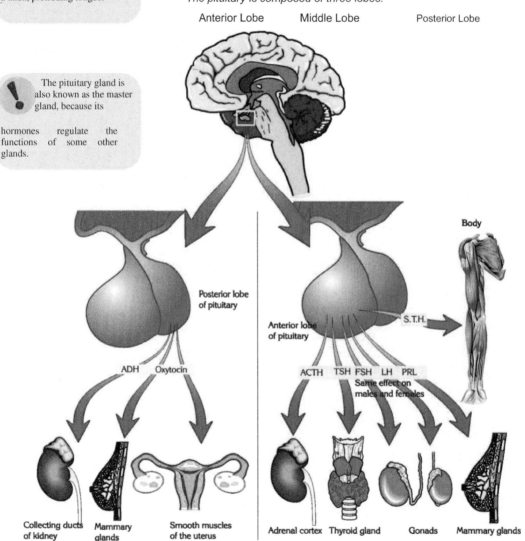

Anterior Lobe   Middle Lobe   Posterior Lobe

1. Anterior Lobe

This lobe constitutes two fifths of the volume of the pituitary and is composed of epithelial tissue. Many hormones are released into the blood by this lobe. The anterior lobe mainly produces six different types of hormones. Four of them are tropic hormones, whose secretions cause other endocrine glands to release their own hormones. Other hormones of this lobe are growth hormone and prolactin.

a. Tropic Hormones

These hormones stimulate the secretion of hormones from other endocrine glands.

Thyroid stimulating hormone (TSH) activates the thyroid gland and causes secretion of the hormone thyroxine.

Adrenocorticotropic hormone (ACTH) stimulates the adrenal glands directly above the kidney and controls the secretion of hormones from these regions.

Gonadotrophic hormones (LH, FSH) stimulate glands in the organs of the reproductive system (gonads).

Follicle stimulating hormone (FSH) is transported to the testes or ovaries via the blood. In the ovaries, growth and maturation of the ova (egg) within the follicle is controlled by this hormone. It simultaneously stimulates the secretion of estrogen from the follicle. In the testes, it stimulates the production of sperm.

Luteinizing Hormone (LH) triggers the release of the ovum by rupture of the follicle. In females, it also stimulates the secretion of progesterone from the corpus luteum, which occurs after ovulation. In males it triggers some special cells in the testes to produce testesterone (male sex hormone).

b. Growth hormone and PRL

Growth hormone is released throughout the lifetime of an individual. During childhood and adulthood, up to approximately 24 years of age, the height of the body increases due to an increase in the length of the bones.

A deficiency in this hormone in childhood will result in a condition known as dwarfism, where those affected are shorter than average. An excess amount of this hormone, also in childhood, causes an abnormal increase in body length. This condition is known as gigantism.

In gigantism, an oversecretion of growth hormone in adulthood causes acromegali, causing parts of the body to grow disproportionaly. Overall body length does not increase, but abnormally large hands, feet, jaw and lips can be seen.

Growth hormone affects protein, carbohydrate and lipid metabolism. It causes an increase in the amount of lipids, protein synthesis and glucose secretion from the liver into the blood. It also increases the rate of cell division.

Prolactin (PRL) is produced only by females. It triggers milk production from the mammary glands.

 Hypothyroidism in adults produces myxedema. As in the cretin, there is dry skin, a thick tongue, lowered mentality or apathy, lowered temperature and basal metabolic rate, anemia and hypercholesterolemia. The face is puffy, and the cheeks are flabby. There is a husky, bass voice due to swelling of the larynx. There is decreased libido or sexual impotency, and, in women, oligomenorrhea, amenorrhea or sterility.

Dwarfism is an endocrine disorder due to insufficient secretion of growth hormone. During the development of a normal individual, the pituitary secretes growth hormone, which stimulates growth of bone and muscle. Pituitary dwarfs undersecrete this hormone, with the result that they fail to increase in height beyond that expected of a 6 year old child. The proportions of their body and their intelligence are normal, however.

Children who are seen to be growing at a greatly reduced rate can be given injections of growth hormone that will reduce the intensity of the disorder. Another type of dwarfism, also known as cretinism, is caused by a thyroid hormone deficiency.

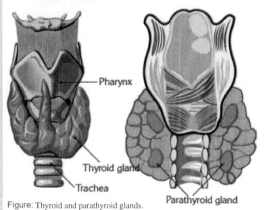

— Pharynx

Thyroid gland

Trachea

Parathyroid gland

Figure: Thyroid and parathyroid glands.

2. The Middle Lobe

The middle lobe is composed of epithelial tissue. It is seen only during the fetal stage of development. From this lobe, MSH (melanin stimulating hormone), which stimulates melanocytes, is secreted. It affects the cells located within the skin containing melanin pigment. MSH determines the color of the skin. It is active in amphibia, reptiles and birds. Its effect is uncertain in mammals.

3. The Posterior Lobe

The posterior lobe is composed of nerve tissue and the following hormones are secreted from this region:

ADH (Antidiuretic hormone)

oxytocin

a. Antidiuretic Hormone (ADH), also known as vasopressin:

The main function of ADH is to regulate the water balance of the body by controlling reabsorption of water in the kidneys. ADH causes an increase in water permeability of the cells of the distal channel of the nephron (main part of the kidneys where filtration of blood and reabsorption of necessary molecules occurs).

The most important agent regulating the secretion of ADH is the concentration of the blood. The hypothalamus contains receptors sensitive to the osmotic potential of the blood. If the osmotic pressure of the blood increases, secretion of ADH subsequently increases. If the osmotic pressure of the blood decreases, ADH secretion also decreases. An increase in the volume of blood decreases ADH secretion. This is detected by stress receptors in veins so that, in the case of a decrease in volume, ADH secretion increases. (Alcohol, caffeine and such substances directly affect ADH secretion. ADH secretion slows. Water is not reabsorbed properly and, within a short time, the urinary bladder becomes full of urine).

b. Oxytocin

This hormone stimulates the smooth muscle fibers of the uterus to contract during labor, and after birth to expel the placenta. Oxytocin hormone can be injected into the body in order to accelerate the birth process. Oxytocin also affects the mammary glands. It causes contractions of the channel cells of these glands so that milk can be secreted. The function of oxytocin in males is unknown.

c. Thyroid Gland

The thyroid gland is located at the front of the neck, directly beneath the voice box. The thyroid gland is found in all vertebrates, and in humans it weighs approximately 25 g. In mammals, the thyroid gland is composed of two lobes. It has the highest capillary content as compared to all the other endocrine glands.

Each lobe is enclosed within a capsule composed of connective tissue. This capsule subdivides the thyroid into smaller lobes, and is composed of secretory vesicles known as follicles. The small lobes of these cells secrete thyroxine. The thyroid is made up of a great number of spherical follicles filled with a gelatin-like liquid. The thyroid hormone thyroxine is stored within this liquid. Thyroxine ($C_{15}H_{11}O_4N_4$) is a hormone with a high iodine and peptide content. This hormone can also be manufactured synthetically.

If the thyroid gland is removed from a young individual, body growth is arrested and the rate of metabolism decreases. In fact, most of the functions of the body are affected, and this may even result in death. These effects are less serious in adults. When thyroxine is given to adult animals, the effects of its deficiency disappear. When the thyroid gland secretes less hormone than normal in adult humans, the normal body temperature decreases by two degrees. In addition, the body accumulates lipids and obesity results, as does acne and loss of hair. The patient's face also swells and there is bagginess under the eyes. These conditions are symptoms of the disease **myxodema** (hypothyroidism in adulthood). Metabolic functions are characteristically considerably lower than normal in these individuals. Normal metabolic function is, however, regained if the thyroid gland is replaced. If the thyroid fails to develp properly in childhood, a condition called cretinism results. Individuals with this condition are short and stocky. Thyroxin therapy can initiate growth, but unless treatment is begun within the first two months, mental retardation results.

On the other hand, if the thyroid produces too much of the thyroid hormones, a person may feel as though the "engine is racing", with such symptoms as a rapid heartbeat, nervousness, weight loss, and protrusion of the eyes. This condition is known as hyperthyroidism. Thyroxine (T4) and Triiodothyronine (T3) are iodine-containing hormones (numbers given here show the amount of iodine in the structure of the hormones). When the diet lacks iodine, not enough iodine will be available, the thyroid gland cannot produce its hormones, and the thyroid swells. This condition is called goiter. Regular consumption of fish, particularly saltwater fish, will provide adequate amounts of iodine. This element is necessary for the normal function of the nervous system.

Another rich source of iodine is garlic. It is found in lesser amounts in lemon, onion and radish. In addition, garlic cleanses the intestines, strengthens the bones and helps to prevent infection. It is believed to have a role in the prevention of cancer. The hormone calcitonin is also secreted from the thyroid gland. Calcitonin, together with parathormone from the parathyroid gland, plays an important role in the regulation of calcium phosphate concentration in the body. Calcitonin decreases the amount of calcium in the blood. It also affects the amount of calcium excreted in the urine.

The synthesis and release of thyroxine from the thyroid gland is regulated by TSH secreted from the anterior lobe of the pituitary. Animal research has proved the effect of the pituitary on the thyroid gland. TSH is transported to the thyroid in the blood. If the concentration of TSH exceeds a centain level, a system of negative feedback prevents further secretion of TRH and TSH. Thus, the level of thyroxine is regulated.

Figure: An **exophthalmic goiter** (hyperthyroidism) appears as a result of an over-active thyroid gland, where the thyroid grows to an abnormal size. Due to the overproduction of thyroxin, an increase in metabolism occurs.

Figure: **Simple goiter** usually occurs when iodine is insufficient in the diet.

 Diabetes is an endocrine disorder that affects the pancreas. There are two types of diabetes. Type I (Insulin dependent)

Symptoms usually appear in childhood. This form of the disorder results when the Beta cells of the pancreas fail to produce adequate amounts of insulin. After a carbohydrate containing meal, the level of blood glucose rises, a situation that should, but fails to, stimulate the production of insulin.

d. Parathyroid Gland

A parathyroid gland is found in all vertebrates, excluding fish. It comprises four distinct areas of tissue embedded in the thyroid gland. Two areas form a pair at the base of the gland, and the other pair is found at the top. Their number may sometimes fluctuate, as may their size, which ranges from that of a pea to that of a bean. When the thyroid gland is removed, these glands are also removed.

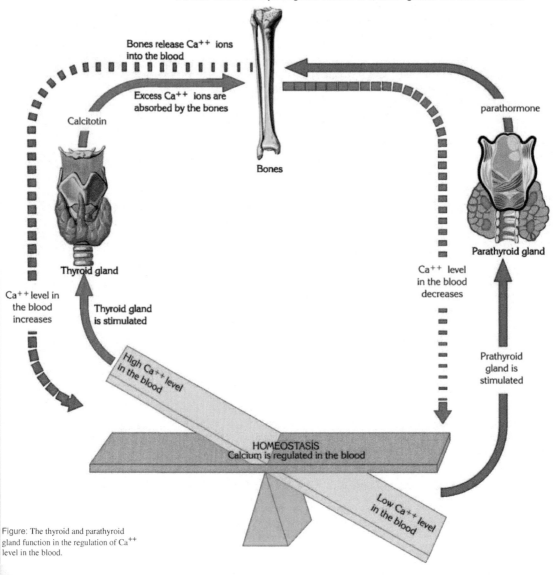

Bones release Ca^{++} ions into the blood

Excess Ca^{++} ions are absorbed by the bones

Calcitotin

Bones

parathormone

Thyroid gland

Parathyroid gland

Ca^{++} level in the blood increases

Ca^{++} level in the blood decreases

Thyroid gland is stimulated

High Ca^{++} level in the blood

Prathyroid gland is stimulated

HOMEOSTASIS
Calcium is regulated in the blood

Low Ca^{++} level in the blood

Figure: The thyroid and parathyroid gland function in the regulation of Ca^{++} level in the blood.

Parathyroid hormone (PTH) produced in these glands regulates the amount of calcium and phosphate in the blood. It causes the blood phosphate (HPO4) level to decrease and the blood calcium (Ca^{++}) level to increase. The exact amount of calcium is necessary for the normal functioning of muscles, heart, skeleton and nerves, and is regulated hormonally by the thyroid and parathyroid, which transfer calcium between the bones and the blood. The secretion of PTH is regulated according to the Ca^{++} concentration in the blood, without the direct effect of the pituitary and nervous system.

When the amount of calcium decreases, this hormone accelerates the passage of Ca^{++} from bone to the blood by directly affecting bone cells. While PTH causes an increase in Ca^{++} concentration in blood, it decreases the amount of phosphate when the Ca^{++} concentration in blood exceeds a certain value. It is then secreted from the thyroid gland. Calcitonin then stimulates the uptake from the blood of excess calcium, which is subsequently stored in the bones

An excessive amount of parathyroid hormone causes the destruction of the muscles, an increase in the fragility of bones and defects in the nervous system. When this hormone is present in reduced amounts, the calcium concentration in the blood decreases and, as a result, the functions of muscles and nerves increase and sudden contractions result, also known as muscle tetany.

e. Pancreas

As explained previously, the pancreas contains both endocrine and exocrine glands. These types of glands are known as compound glands. Endocrine secretion in the pancreas is performed by the islets of Langerhans. This gland produces and secretes two well-known hormones: insulin and glucagon.

Insulin and glucagon are effective in carbohydrate, lipid and protein metabolism. The main function of insulin is to promote the penetration of glucose into tissue cells from the blood by increasing the permeability of the cell membrane. It provides glucose as an energy source when required and also stores glucose as glycogen in the liver and muscle cells in a process known as glycogenesis.

It accelerates the passage of glucose into adipose tissue and its conversion to fatty acids. It reduces protein metabolism and causes an increase in protein synthesis. Simply, insulin decreases blood glucose concentrations by either causing cells to use it as an energy source or change it to glycogen, a much less soluble form of carbohydrate, and storing it in the liver and muscle cells, to be used later.

A lack of insulin is known to cause diabetes mellitus. There are two main types of this disorder. In type I (insulin dependent) diabetes, the pancreas does not produce insulin. The condition is believed to be brought on by exposure to an enviromental agent, most likely a virus, whose presence causes cytotoxic T cells to destroy the pancreatic islets of Langerhans (an autoimmune disease). Not having the ability to produce insulin may lead to hyperglycemia. As a result, sugar concentration in the blood increases, as well as the osmotic pressure of the blood

 As glucose accumulates in the blood, it is also excreted in the urine.

Other symptoms of type-I diabetes include thirst, sweating and the excretion of large amounts of urine.

Sufferers have to take an injection of insulin before each meal and regulate their intake of carbohydrate. Those who have injected too much insulin into their body or have missed a meal can become **hypoglycemic,** and need to keep glucose available to prevent their blood sugar from dropping to critical levels.

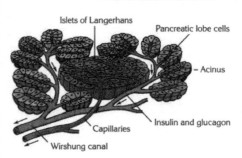

Figure: The cells of the islets of Langerhans are responsible for insulin and glucagon production.

Type II Diabetes (Insulin independent)

Symptoms usually appear in adults, particularly those who are overweight. In contrast to type I diabetes, sufferers produce enough insulin. The cause of the condition is resistance of the target cells to insulin. As is the case for sufferers of type I, regular medication is necessary; obviously not insulin, but oral hypoglycemic agents.

Those suffering from Type II diabetes should carefully control their diet, avoiding excessive carbohydrate-containing foods.

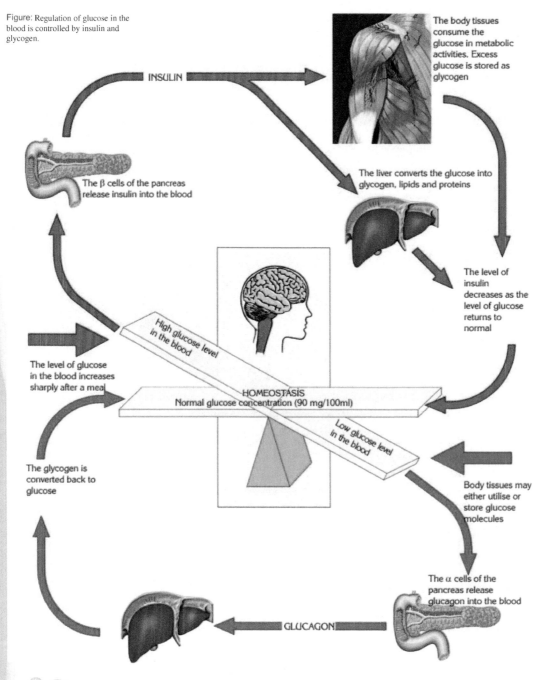

Figure: Regulation of glucose in the blood is controlled by insulin and glycogen.

INSULIN

The body tissues consume the glucose in metabolic activities. Excess glucose is stored as glycogen

The β cells of the pancreas release insulin into the blood

The liver converts the glucose into glycogen, lipids and proteins

The level of insulin decreases as the level of glucose returns to normal

High glucose level in the blood

The level of glucose in the blood increases sharply after a meal

HOMEOSTASIS
Normal glucose concentration (90 mg/100ml)

Low glucose level in the blood

The glycogen is converted back to glucose

Body tissues may either utilise or store glucose molecules

The α cells of the pancreas release glucagon into the blood

GLUCAGON

This condition causes the excretion of an excessive amount of glucose in the urine, together with a large volume of water. Cells are dependent on lipids and proteins as an energy source when there is insufficient glucose available due to a lack of insulin. As protein synthesis decreases, deficiency symptoms then start to appear in the body. Cessation of lipid synthesis, however, causes an increase in fatty acid catabolism and there is an increase in ketose sugars and acidity, also known as acidosis. An increase in the osmotic pressure of the blood, loss of water and the toxic effect of ketosis on the nervous system caused by an excess of ketose sugars triggers coma in diabetics if insulin is not administered promptly. The individual may die from this condition.

To prevents these symptoms, the individual must have daily insulin injections. The injections control the diabetic symptoms, but still can cause some problems, since either an overdose of insulin or the absence of regular eating can bring on the symptoms of hypoglycemia (low blood sugar). In a such case, the cure is quite simple: a sugar cube or fruit juice can very quickly counteract hypoglycemia.

Probably more than 90% of of diabetic people have type II (noninsulin-dependent) diabetes. This type of diabetes occurs in people of any age who are obese and inactive (usually after age forty).

The pancreas either produces insufficent insulin, or the cells do not respond to it, mainly because of not having functional receptors necessary to detect the presence of insulin, and, therefore, the liver and the muscles in particular are incapable of taking up glucose. These people can often treat their disorder with only diet and exercise. If type II diabetes is untreated, the results can be as serious as type I diabetes (Diabetics are prone to blindness, kidney disease, and circulatory problems).

Glucagon, a second hormone secreted by the islets of Langerhans, enables the concentration of blood sugar to increase by increasing the rate of glycogen catabolism in the liver. When the blood glucose level is low, glucagon triggers the liver to convert glycogen to glucose. As the concentration of glucose increases, glucagon secretion is reduced. The presence of insulin acts as an inhibitor.

Secretions from the pancreas are regulated by feedback from the concentration of glucose in the blood passing through the pancreas. An increase in glucose concentration in the blood triggers an increase in insulin secretion. In contrast, a decrease in glucose concentration results in an increase in glucagon secretion.

f. Adrenal Glands

The human body has two adrenal glands, each positioned on top of a kidney. Each one is approximately 12 g in weight and is characterized by its rich capillary network. The adrenal glands have a more extensive capillary network as compared to other organs of the body. In adults, the adrenal glands have fibrillar contact with the kidneys and, if a kidney is removed, its neighboring adrenal gland remains unaffected.

The adrenal glands are composed of two distinct regions. The outer, light-yellow region is known as the adrenal cortex, and the inne,r dark-brown region is known as the adrenal medulla.

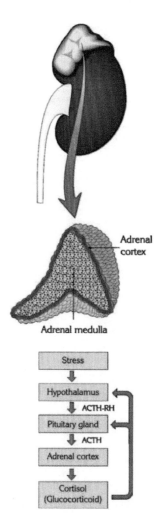

Figure: The structure and function of an adrenal gland. The medulla and cortex of the adrenal glands function separately in the production of different hormones.

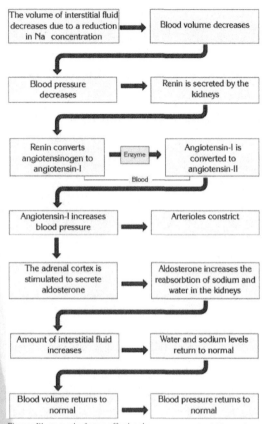

The volume of interstitial fluid decreases due to a reduction in Na concentration → Blood volume decreases

Blood pressure decreases → Renin is secreted by the kidneys

Renin converts angiotensinogen to angiotensin-I → [Enzyme] → Angiotensin-I is converted to angiotensin-II

Blood

Angiotensin-I increases blood pressure → Arterioles constrict

The adrenal cortex is stimulated to secrete aldosterone → Aldosterone increases the reabsorbtion of sodium and water in the kidneys

Amount of interstitial fluid increases → Water and sodium levels return to normal

Blood volume returns to normal → Blood pressure returns to normal

Figure: Illustrates the factors affecting the regulation of blood pressure.

Over activity of the adrenal cortex in females results in masculinization, facial hair formation, developing of testes, deepening of voice and reduction in the size of uterus and ovaries.

1. Adrenal Cortex

The adrenal cortex constitutes the most important part of the gland, since it is the site of hormone secretion. These hormones are known as corticoids, and are steroidal in structure. Adrenal cortical hormones are divided into three groups according to their functions.

Glucocorticoids

Mineralocorticoids

Sex hormones.

a. Glucocorticoids

These hormones affect carbohydrate, lipid and protein metabolism. They induce production of cortisone, cortisol and corticosteroid hormones. The most important of these is cortisol. This hormone increases the level of blood sugar. It promotes the synthesis of glucose from noncarbohydrates such as lipids and proteins. In addition, it prevents oxidation of glucose and causes an increase in the amount of nitrogen in the urine.

It also causes an increase in amino acid and protein digestion. These hormones play an active role when the body is under stress and maintain the homeostatic balance of the body. They are responsible for the provision and regulation of energy sources under varying conditions. Glucose secretion increases under different stressful conditions. Adrenocorticotropic hormone (ACTH) secreted by the pituitary regulates the secretion of these hormones.

b. Mineralocorticoids

Mineralocorticoids are hormones that regulate the mineral balance of the body. The most important member of this group is aldosterone. Aldosterone accounts for 95% of mineralocorticoid function. Aldosterone and other mineralocorticoids affect the reabsorption of Na^+ (returning Na^+ back to the bloodstream) from the kidneys. Over secretion of aldosterone causes an increase in Na^+ reabsorption from the kidneys. Its deficiency also leads to an increase in K^+ loss in urine. A long term lack of K^+ causes kidney disorders and muscle wasting. The amount of Na^+ and Cl^- in the blood decreases in the absence of mineralocorticoid hormones.

Aldosterone secretion can be stimulated by ACTH. Secretion however is generally regulated by the renin-angiotensin system through a feedback mechanism. The amount of renin excreted into the blood by the juxtaglomerular cells of the kidneys increases as the blood volume decreases, or as the amount of Na^+ decreases. When the blood volume decreases, the angiotensin–aldosterone system maintains the blood volume at a specific level by the repetition of this cycle.

c. Sex Hormones

Sex hormones are secreted by the adrenal cortex of both males and females. The most effective androgenous hormone, testosterone, is secreted by the testes and only a small amount is secreted by the adrenal cortex.

Over activity of the adrenal cortex in a male child results in an increase in androgenous hormone production which triggers the onset of puberty earlier than normal. Overactivity of the adrenal cortex in females results in masculinization. Facial hair starts to grow, testes develop, the voice deepens and the ovaries and uterus decrease in size.

The adrenal glands, placenta and ovaries have a higher concentration of cholesterol than other organs since all steroid hormones are synthesised from Acetyl-CoA and cholesterol.

When there is a low level of adrenal cortex hormones due to hyposecretion, a person develops Addison disease, characterized by low blood glucose, weight loss, weakness and peculiar bronzing of the skin.

Left untreated, Addison disease can be fatal. When there is a high level of glucocorticoids (cortisol), a person develops Cushing syndrome. The person usually has an obese trunk and "moon face" while the arms and legs remain normal. Such a person also has high blood glucose and Na levels because of excess cortisol.

2. Adrenal Medulla

The adrenal medulla is different from most other endocrine tissue in that its cells are derived from cells of the peripherial nervous system and are specialized to secrete hormones. Hormones of the adrenal medulla are epinephrine (adrenaline) and norepinephrine (noradrenaline). These hormones respond to stress. They are involved in the immediate response to stress. They bring about all bodily changes that occur when an individual reacts to an emergency. Heart and breathing rates and blood flow increase. If a person continues to be highly stressed over a long time, the result can even be death.

g. Gonads

The main function of the male and female gonads is in the development of the reproductive system. They also function as endocrine glands. The hormones secreted by the gonads are steroid in structure. LH and FSH hormones stimulate hormone secretion from the ovaries and testes. In the presence of these hormones, estrogen and progesterone are secreted by the ovaries and testosterone is secreted by the testes.

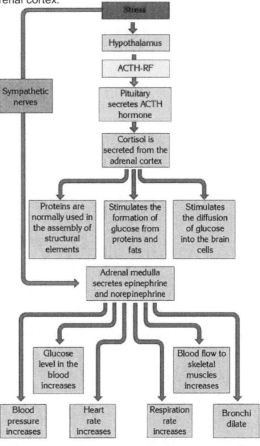

Figure: Illustrates the response of the body to a stressful stimulus.

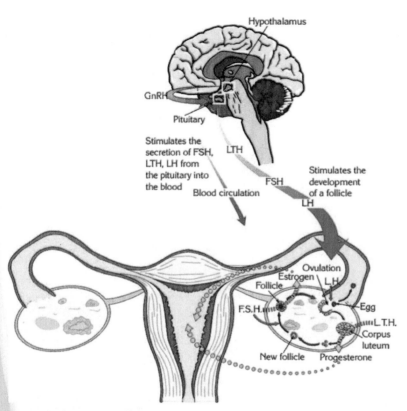

Figure: GnRH produced by the hypothalamus stimulates the secretion of FSH, LTH and LH from the pituitary into the blood. FSH promotes the development of a follicle. It releases estrogen which thickens the endometrial lining of the uterus to receive a fertilized egg. As the level of estrogen rises, LH triggers ovulation from the follicle.

After the ovum is released, the burst follicle, now known as the corpus luteum, secretes estrogen and progesterone, which maintain the uterine lining. If no fertilization occurs, estrogen and progesterone secretion cease, triggering menstruation.

1. Ovaries

The hormones estrogen and progesterone are secreted by the ovaries.

a. Estrogen

Estrogen is secreted by follicular cells of the ovaries in response to stimulation by FSH. Its action results in thickening of the uterine lining. It also affects female secondary sexual characteristics. During pregnancy, estrogen is secreted by the placenta .

b. Progesterone

Progesterone is secreted by the follicle, also known as the **corpus luteum**. Its secretion is the result of the rupture of the follicle due to ovum release under the stimulus of LH. Progesterone stimulates the uterine mucosa to thicken in preparation for implantation of a fertilized egg.

Progesterone maintains pregnancy by causing the muscle cells of the uterus to relax. When an ovum has been fertilized, further ovulation ceases.

Figure: Synthetic testosterone used for body building may cause severe side effects.

Progesterone also affects the development of the mammary glands. Progesterone, together with estrogen, plays an important role in regulation of the menstrual cycle. The corpus luteum continues to secrete progesterone until the fourth month of pregnancy. This function is then performed by the placenta.

2. Testes

Male sex hormone, testosterone, is produced by the testes under the influence of luteinizing hormone (LH) and follicle stimulating hormone (FSH). Testosterone functions with LH and FSH to stimulate the production of sperm. It is involved in growth and maintenance of male sex organs and secondary sex characteristics, including facial hair, deepining of the voice, and increased muscle growth.

h. Thymus Gland

The thymus gland is located in the thorax, a few centimeters below the thyroid. The thymus produces a variety of hormones to promote development of certain immune cells called T lymphocytes. This gland is quite active during childhood, but is replaced by fat and connective tissue and becomes non-functional in adulthood.

I. Epiphysis (The Pineal Gland)

The epiphysis is located between the hemispheres of the brain and protrudes onto the upper surface of the hypothalamus.

Melatonin is secreted by the epiphysis. Melatonin prevents LH and FSH hormone secretion by the pituitary before puberty.

The epiphysis prevents early onset of puberty by affecting the functions of the pituitary and the gonads during infancy.

A decrease in the concentration of melatonin is seen during puberty, and this leads to an increase in FSH and LH plasma concentration. Melatonin production depends upon the pattern of light and dark an individual is exposed to. Light inhibits melatonin synthesis; darkness stimulates it.

Changing levels of the hormone may signal an organism to prepare for changing seasons. The role of the human pineal gland is not well understood, possibly because we alter natural light/dark cycles with artificial lighting.

j. Nonendocrine hormones

The principle endocrine glands and their secretions have been mentioned up to now. However, certain cells in the body produce chemical messengers that regulate nearby cells without travelling in the blood stream and are therefore not considered endocrine hormones.

These intercellular chemical messengers are often called local hormones, or prostaglandins. Chemicals produced by the uterus cause contraction at birth. Some others that are produced by the stomach and intestines help digestion to occur properly. Enkephalins and endorphins are local brain hormones which have pain-killing properties similar to the drug morphine.

Anabolic Steroids

There are not to be confused with the various corticosteroid drugs often prescribed by doctors for conditions such as rheumatoid arthritis and Crohn's disease. Anabolic steroids have only a very limited use in ordinary medicine, such as building up muscle in a patient who has been bedridden for some time.

They are misused by some athletes and bodybuilders in order to increase muscle size and strength. Because they are derived from the male hormone testosterone, they may also have the effect of stimulating aggression. Some athletes claim the drugs help them train harder and recover more quickly from injury. Their use is prohibited by sporting bodies, and athletes found using steroids are liable to be disqualified and banned from competing.

There are many potential problems associated with these drugs:

often steroids are injected using shared equipment, with the attendant risks of HIV and other infections;

in young people, steroids can restrict growth;

in men, they have side-ef-fects on the reproductive system, such as reduced sex drive and lowered sperm count;

in women, 'masculine' side-effects such as deeper voice and smaller breasts may be permanent even when the drug use has stopped;

they can cause liver damage, leading to jaundice or cancer of the liver.

These drugs circulate in some public gyms and health clubs. It is illegal to supply them without a prescription, but not illegal to possess them.

ENDOCRINE GLANDS - HORMONES & THEIR EFFECTS

GLAND			HORMONE	EFFECTS
Pituitary	Posterior lobe of the pituitary		Antidiuretic hormone (ADH) (Vasopressin)	It stimulates the contraction of smooth muscles, increases blood pressure and increases reabsorption of water from the kidneys.
			Oxytocin	It stimulates milk secretion and contraction of uterine muscles.
	Anterior lobe of the pituitary		Growth hormone (Somatotropic hormone = STH)	It controls growth and bone formation since it affects carbohydrate, lipid and protein metabolism.
		Gona dotro pins	Prolactin (LTH)	It promotes development of mammary glands during pregnancy and then regulates the production of milk after birth. It also initiates the mothering instinct.
			Luteinizing hormone (LH)	It is involved in release of the ovum from the follicle and release of progesterone from the corpus luteum in females. It triggers the secretion of testosterone in males.
			Follicle stimulating hormone (FSH)	It stimulates the development of an ovum from one of the ovaries. It also stimulates the secretion of estrogen from the developing follicles.
			Adreno-corticotropic hormone (ACTH)	It stimulates the secretion of cortisol from the adrenal cortex.
			Thyroid-Stimulating hormone (TSH)	It stimulates the secretion of thyroxine from the thyroid gland.
			Melanocyte stimulating hormone (MSH)	It stimulates melanin production from melanocytes in the skin.
Thyroid gland			Thyroxine	It accelerates metabolism and heart rate.
			Calcitonin	It reduces the concentration of calcium in the blood.
Parathyroid gland			Parathormone	It increases the concentration of calcium in the blood.
Adrenal glands	Cortex		Cortisol	It influences carbohydrate, fat and protein metabolism. It is mainly involved in the conversion of proteins to carbohydrates.
			Aldosterone	It maintains homeostasis of sodium in the blood by regulating the concentration of sodium reabsorbed in the loop of Henlé. It is also involved in the excretion of excess potassium.
	Medulla		Epinephrine (Adrenaline)	They increase blood pressure, glucose level and flow rate.
			Norepinephrine	
Pancreas (islets of Langerhans)	β cells		Insulin	It maintains glucose homeostasis in the blood by converting any excess into insoluble glycogen..
	α cells		Glucagon	It maintains glucose homeostasis in the blood by converting glycogen into glucose.
Ovarium	Follicle		Estrogen	It stimulates formation of secondary sexual characteristics, maturation of reproductive structures, and thickening of the endometrium.
	Corpus luteum		Progesterone	It stimulates the development of mammary glands. It maintains the endometrium (lining of uterus) during pregnancy.
Testes			Testesterone	It stimulates the formation of secondary sex characteristics..
Placenta			Estrogen, progesterone	It is secreted during embryonic development.

SELF CHECK

EndocrineSystem

A. Key Terms

Aldosterone	Diabetes mellitus
Glucagon	Hormone
Hyperthyroidism	Islets of Langerhans
Negative feedback	Steroid hormones
Target cell	Thyroxine
Cretinism	Dwarfism

B. Review Questions

1. Explain why hormones affect only their target organs.

2. Compare the mechanism of steroid and peptide hormones on cell metabolism.

3. Give examples of and explain the hormone-related disorders of the body.

4. How is a hormone deficiency of the body treated?

5. Explain the feedback mechanism, using glucose level regulation in the body.

6. Explain the source and function of five hormones secreted in the human body.

7. List the sex-related hormones secreted by the hypothalamus and the pituitary.

8. Give the names of all hormones secreted by the hypothalamus and the pituitary.

9. Pancreas → insulin, glucagon Adrenal gland → Cortisol, Aldosterone Pituitary gland → ACTH, FSH Ovary → Estrogen, progesterone

Various glands and their hormones are listed above. Simply explain the function of each hormone

10. - Myxodema

 - Dwarfism

 - Goiter

 - Cretinism

Explain the causes of the above hormone related disorders.

C. True or False

1. Peptide hormones are received by receptors of the cell membranes.

2. Hypothalamus produces neurohormones.

3. Because it controls some other glands, the pituitary is known as the master gland.

4. When growth hormone is secreted more than normal in adulthood, it causes acromegaly.

5. Enlargement of the thyroid gland is due to not having enough Na^+ in the diet.

D. Matching

a. Myxodema ( ) Regular physiological or behavioral event that occurs on an approximately 24 hour cycle.

b. Prolactin ( ) Hormones that have various and powerful local effects.

c. Hormone ( ) Hormone of the anterior pituitary that stimulates production of milk from the mammary gland.

d. Circadian rhythm ( ) Condition resulting from a deficiency of thyroid hormone (hypothyroidism) in an adult.

e. Prostaglandins ( ) Chemical messenger that has physiological and developmental effects, usually on another part of the organism.

Sex hormones affect development of brain. Nervous system controls the endocrine system via the hypothalamus.

Many hormones work together to control water and mineral reobsorbtion and secretion by kidneys.

Kidneys keep blood concentration at a certain level which aids hormone transportation.

System

Skeletal

Parathormone, and calcitonin regulates minaral concentration of the skeleton. Grawth hormone regulates bone growth. Some minerals which can be used by the glands are stored in the bones.

System

Muscular

Adr
mus
pow
hor
test
mus
cover and protect the glands.

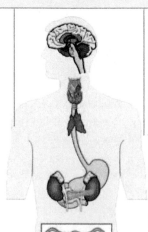

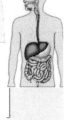

CO$_2$ from the body.

Blood volume is regulated by the synergic activities of several hormones.

Blood delivers hormones to their targets. Atrial natriuretric hormone is produced by the heart to regulate blood volume.

Locomotion Systems

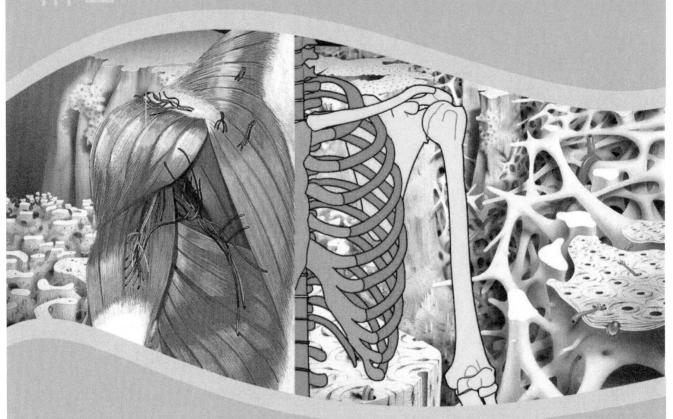

LOCOMOTION SYSTEMS

An organism's ability to respond to stimuli is important for homeostasis. Movement of the body, any time needed, is a remarkable activity to keep organisms in homeostasis. Organisms provide this important ability by means of two systems. The skeletal system and the muscular system. The term locomotion systems refers to these two systems together.

Skeletal system: Besides being used in movement, the skeleton is a supporting structure or framework. It gives body shape to organisms and protects their internal organs. Several types of skeletons perform these functions in different organisms.

Hydrostatic skeleton: This is the simplest type of skeleton ("hydro" means water). It consists of liquid within a layer of flexible tissue, as in some invertebrates, such as worms, and simple plants

Exoskeleton: It consists of solid structures, strong enough to resist pulling forces. Muscles can attach to the surface of the framework, facilitating movement. An exoskeleton protects organisms from the outside. Many groups of invertebrates have exoskeletons, such as insects and some mollusks (snails). An exoskeleton is made of either chitin, a complex carbohydrate, or $CaCO_3$. Organisms with a chitin exoskeleton may change their exoskeleton a few times during their life in order to grow, since the exoskeleton prevents growth. Changing the chitin exoskeleton is known as ecidysis, or molting. Organisms with a $CaCO_3$ exoskeleton add new rings to their exoskeleton when they grow. All types of exoskeleton are produced by the organism's living cells, but the exoskeleton is not a living part of the organism.

Endoskeleton: The vertebrate endoskeleton has several functions

Movement: This is done by the cooperative action of the muscles and bones. Muscles attach to the bones and pull them when they contract. This activity causes movement.

Support and giving shape: The skeleton is a strong framework which gives a certain shape to the body. The body would collapse into a shapeless heap without the skeleton.

Protection of the internal organs: Some bones form the rib cage, where the heart and lungs are protected. The vertebral column protects the spinal cord, and the brain is protected by another bony structure, the skull

Production of blood cells: Many bones, such as the long bones of the human arms and legs, contain red bone marrow, a tissue which produces blood cells.

Mineral storage: The skeleton stores some minerals, such as calcium and phosphorus. These minerals give a certain strength to the skeleton and also are used in many metabolic activities, as needed. Of the human body's calcium, 99% is stored in the bones.

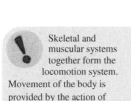

Skeletal and muscular systems together form the locomotion system. Movement of the body is provided by the action of these two systems.

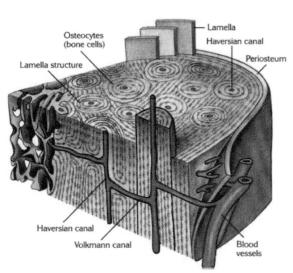

Osteocytes (bone cells)

Lamella structure

Lamella

Haversian canal

Periosteum

Haversian canal

Volkmann canal

Blood vessels

Figure: A cut-away diagramatic view of bone, illustrating its structures.

The bones of the vertebrate skeleton are grouped into two parts. The **axial skeleton**, so named because it is located in the longitudinal central axis of the body, consists of the skull, vertebral column, ribs. and sternum (breastbone).

The **appendicular skeleton** consists of the limbs (arms with hands, legs with feet) and girdles (pectoral girdle--clavicle and scapula, which support the forelimbs; and the pelvic girdle, which support the hindlimbs).

HUMAN SKELETAL SYSTEM

In an adult human, 206 bones make up the skeletal system (This number is more than 300 in babies). Actually, the skeleton is composed of cartilage and bone developed from connective tissue. Within the bone tissue are bone cells and intercellular matrix secreted by these cells.

These oval flat cells with cytoplasmic projections are known as bone cells, or **osteocytes**. The osteocytes are interconnected by these projections and are located in the lacunae of the bone. Of the intercellular matrix, 45% is minerals, including calcium, phosphate and carbonate, which strengthen the bone. Water constitutes 25% of bone tissue. The intercellular matrix includes a proteinaceous organic compound known as ossein.

The **red bone marrow** is the site of blood cell production. The marrow of long bones is surrounded by compact bone tissue. The tips of long bones and the entire structure of short bones are composed of spongy bone tissue. Compact bone differs from spongy bone tissue in that it contains yellow bone marrow, and there are no spaces within its structure.

Figure: Milk is a rich source of calcium and strengthens bones.

Channels in the ossein allow the entrance and exit of blood vessels and nerves. The parallel main channels within the shaft of the bone are known as **Haversian canals,** and are laterally interconnected by **Volkmann canals**. Circular lamellae radiate from each Haversian canal. Each lamella is bordered by a layer of osteocytes.

1. Bone Formation and Its Regulation

Bones originate from connective tissue. The formation and degradation of bone continues throughout the life span of an individual and is a continuous process. During growth, the rate of formation exceeds that of degradation, resulting in longitudinal extension of bones.

Conversely, after middle age, the rate of degradation exceeds that of formation. During adult life, bone is constantly being broken down by bone-absorbing cells, osteoclasts, which break down bone, remove worn cells and deposit calcium in the blood. The bone is rebuilt (repaired) by another type of cell, the osteoblast.

As they form new bone, osteoblasts take calcium from the blood. Eventually, some of these cells get caught in the matrix they secrete and are converted to real bone cells, osteocytes. There is a continuous exchange of calcium and phosphate between blood and bone tissue.

The quantity of vitamins and minerals in the diet, hormones and genetic factors all influence the concentration of these materials in the bones and in the blood

Locomotion Systems

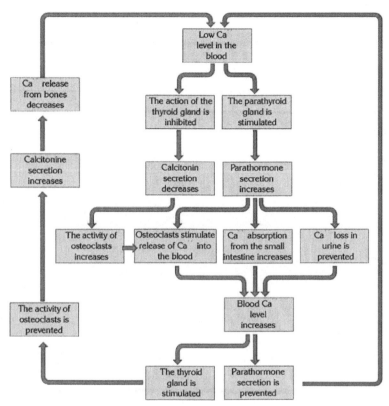

Figure: The steps involved in Ca^{++} regulation in the blood.

Calcium and phosphate are continuously lost via the alimentary tract and the kidneys. Other tissues also require calcium and phosphate and need a constant supply.

If insufficient calcium and phosphate is supplied to the tissues, the remainder is taken from the bones. In long term starvation, the bones donate 1/3 of their mineral reserve to the blood and other tissues to satisfy their needs. Consequently, the bones become soft and fragile.

During pregnancy, the mineral requirements of the fetal skeleton is supplied by the mother. Pregnant women should therefore supplement their diet with either calcium tablets or calcium-rich foods.

In the formation of bone, vitamins A, C and D play important roles. When vitamin D is lacking, calcium and phosphate absorption decreases, consequently teeth and bones take considerable time to form and fail to harden completely. As a result, symptoms of the disease rickets appear. With vitamin A deficiency, the growth rate of the bones decreases. Deficiency of vitamin C causes weak and fragile bones, as well as bleeding gums and abnormally shaped teeth (scurvy).

The formation of bone is regulated antagonistically by parathormone, secreted by the parathyroid gland, and calcitonin, secreted by the thyroid gland. Parathormone is secreted when the level of calcium in the blood decreases, promoting the absorption of calcium from bone to the blood.

Furthermore, it reduces the quantity of calcium lost from the kidneys during excretion and facilitates its absorption from the intestine. The growth and development of the skeleton is regulated indirectly by the effects of somatotrophin (STH--growth hormone) on protein and carbohydrate metabolism. If more STH is secreted than normal, the skeletal system is affected, resulting in gigantism, whereas deficiency of this hormone results in dwarfism.

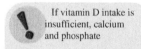 If vitamin D intake is insufficient, calcium and phosphate absorption decreases. Consequently, teeth and bones take considerable time to form and fail to harden completely. This disorder is **rickets**.

2. Types of Bones

There are three types of bone in the human skeleton: long bone, short bone and flat bone.

a. Long Bones

The long bones of the legs are the femur, fibula and tibia. Those of the arms are the humerus, radius and ulna. Each end of a long bone is composed of a wide, expanded region called the **head**. The long part between each head is known as the **shaft**. The external surface of a long bone is covered by a thin membrane known as the periosteum. It functions to increase the diameter of the bone. Compact bone surrounds the head, and spongy bone, containing red bone marrow, forms its center. The shaft of the bone is completely composed of compact bone. At the center of the shaft is a large, longitudinal space filled with yellow bone marrow.

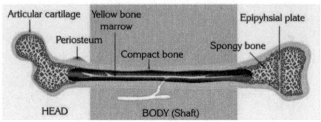

Figure: The structural components of long bone seen in longitudinal section.

b. Flat Bones

The sternum, rib cage, hip bones, patella of the knee joint and the bones of the skull are all flat bones. The layers from the external surface to the center are the periosteum, compact bone, spongy bone and red bone marrow.

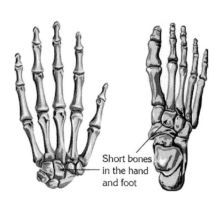

Short bones in the hand and foot

Figure: Bones of hands and feet are short bones.

Figure: Cranial bones are flat bones.

c. Short Bones

They include vertebrae, bones of the hand, foot, wrist joint, ankle joint and fingers. Their structure resembles that of flat bones.

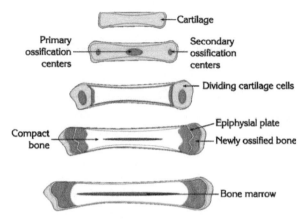

Cartilage

Primary ossification centers — Secondary ossification centers

Dividing cartilage cells

Epiphysial plate

Compact bone — Newly ossified bone

Bone marrow

Figure: The stages in longitudinal growth of long bone.

3. Bone Growth

a. Longitudinal bone growth

The cartilaginous skeleton of the fetus begins to ossify (change to bone) after birth. There are certain ossification centers within the cartilaginous structure, and the cartilage matrix is enriched by minerals. Blood vessels are not found in cartilage but, with the initiation of ossification, they start to penetrate the tissue.

Both ossification and growth of cartilage tissue continues together. The newly formed cartilage tissue ossifies and the bone grows longitudinally as new cartilage is produced. Simply, it is possible to say that cartilage cells (chondrocytes) found in the epiphysal plate are responsible for increasing length. These cells secrete ossein to complete ossification. These processes continue until the age of 17-19. After this age, the epiphysial plates narrow to a thin layer and disappear completely when growth stops at around the age of 23. By this time, the bone is completely formed and composed of compact bone tissue, inside of which is spongy bone tissue. The tips of the bone are covered by **articular cartilage,** which provides a friction-free surface for the smooth articulation of joints.

b. Increase in the diameter of bone

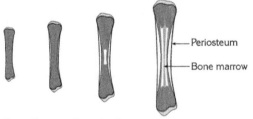

Periosteum

Bone marrow

Figure: The stages of increasing diameter.

All bones increase in diameter by means of the periosteum, a membrane which surrounds them. In juveniles, the new osteocytes diffuse and accumulate to form a new layer under the periosteum. These layers increase the diameter of bone. The periosteum is extremely active until the age of 20, after which its activity slows. The diffusion of osteocytes into the bone was proven by experiment as early as 1840.

c. Fracture Repair

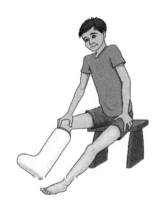

Cartilage helps heal a broken bone. The immediate reaction to a fracture is bleeding, followed rapidly by blood clot formation (hematoma formation). Then, dense connective tissue replaces the blood clot. Cartilage cells enter the connective tissue to form a structure, fibrocartilage callus, to fill the gap the injury left (fibrocartilagenous callus). New spongy bone replaces the cartilage, closing the gap (bony callus). Then the bone hardens, compact bone is formed and the healing process is complete.

Main steps in bone repair:

Bleeding and blood clot formation

Replacement of blood clot with connective tissue

Entrance of cartilage cells to connective tissue

Replacement of spongy bone with cartilage

Formation of compact (hard) bone

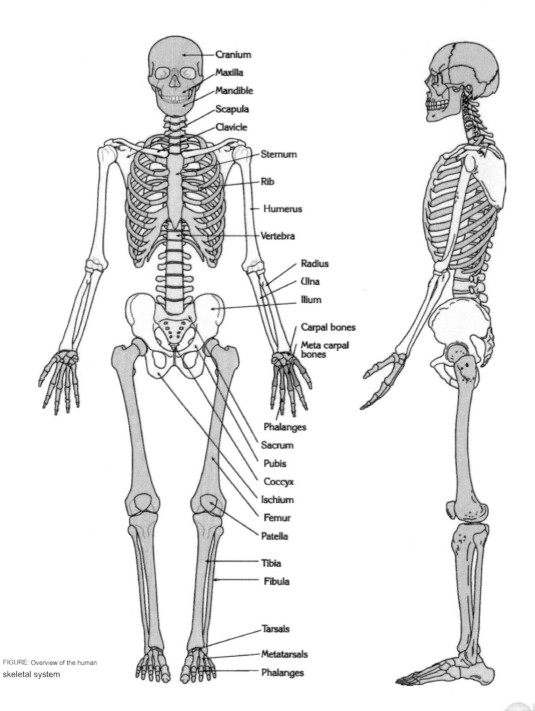

Cranium
Maxilla
Mandible
Scapula
Clavicle
Sternum
Rib
Humerus
Vertebra
Radius
Ulna
Ilium
Carpal bones
Meta carpal bones
Phalanges
Sacrum
Pubis
Coccyx
Ischium
Femur
Patella
Tibia
Fibula
Tarsals
Metatarsals
Phalanges

FIGURE: Overview of the human skeletal system

Locomotion Systems

SKULL (22)			
Facial bones	*(14)*	*Cranial bones*	*(8)*
Lacrimal bones	(2)	Frontal bone	(1)
Zygomatic bones	(2)	Sphenoid bone	(1)
Nasal bones	(2)	Occipital bone	(1)
Nasal concha	(2)	Ethmoid bone	(1)
Maxilla	(2)	Parietal bone	(2)
Paratina bones	(2)	Temporal bone	(2)
Vomer	(1)		
Mandible	(1)		

Cranial bones:

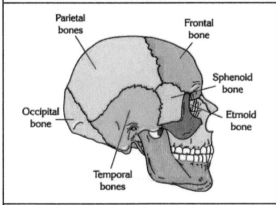

Facial bones:

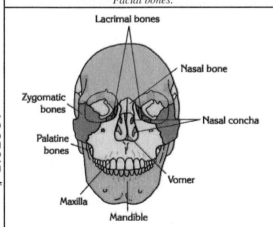

4. Parts of the Human Skeleton

The human body is composed of 206 bones. This number may vary, however, according to age. The skeleton is divided into three main parts: the skull, thorax (trunk) and extremities.

a. Skull

It is composed of cranial bones and facial bones.

1 Cranial bones:

They are fused firmly to each other by immovable joints and enclose important nerve centers such as the cerebrum and cerebellum. During the first year of life, the bones at the side of the skull are very soft and pliable. As the growth process continues, they become first cartilaginous, then ossified. The cranium consists of eight bones: two paired bones, namely parietal bones and temporal bones, and four single bones: the frontal bone, occipital bone, sphenoid bone and the ethmoid bone

2. Facial bones:

They are composed of bones which surround the eyesphere, nasal cavity and mouth; namely, a pair of maxilla, a mandible, the palatine bone, zygomatic bone, lacrimal bone, nasal bone and nasal concha.

b. Trunk

The bones of the trunk are composed of the vertebral column, sternum, ribs, pelvic girdle and pectoral girdle.

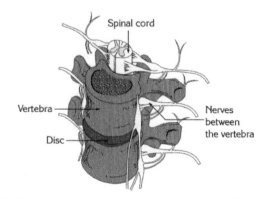

Figure: Vertebrae are stacked directly above each other in a column. The spinal nerves are housed within the vertebral foramina which form an enclosed, protected space running the entire length of the vertebral column

1. The vertebral column

It is the most important supporting structure of the human skeleton, consisting of 33 vertebrae that provide a flexible, protective case for the spinal cord.

The vertebral column has a characteristic "S" shape and is 70-75 cm in length. A vertebral column has five main regions of vertebrae specialized for different functions in various regions of the spinal column:

cervical   thoracic   lumbar

sacral   coccyx

All vertebrae are involved in axial movement of the backbone. The atlas and axis (first two vertebrae) are involved in movements of the head.

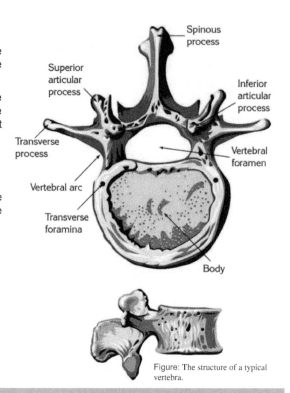

Figure: The structure of a typical vertebra.

Figure: Connection of atlas and axis (left). Posterior view of atlas (right).

READ ME

The most mobile parts of the spinal column are in the lower back and the neck, and these are the most common sites of muscle strain or sprain of the ligaments. In these areas, damage to the intervertebral discs (known as prolapsed or "slipped" discs) sometimes causes irritation or pressure on the adjoining nerve roots or spinal cord. Back and neck strain may occur after prolonged bending (e.g., when gardening), with sudden lifting from a stooped position, or as a result of a "whiplash" injury. Other causes of backache include kidney disease and menstrual disorders.

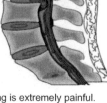

Symptoms and Signs

Dull or severe pain in the lower back (lumbago) or in the neck.

Possible local tenderness.

A spasm of the muscles may occur so that the spine is held rigidly and any attempt at bending is extremely painful.

Pain may pass down the back of the thigh to the lower leg (sciatica), sometimes accompanied by tingling or numbness. If the neck is affected, these symptoms may be felt down the upper limbs.

Aim: Relieve pain and seek medical aid if necessary.

Treatment

Lay the casualty down in the most comfortable position, either on the ground or on a firm mattress, until the pain eases.

If neck pain is severe, fit a neck (cervical) collar to give relief.

If the symptoms persist, seek medical aid.

Locomotion Systems

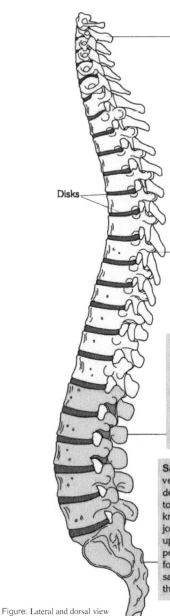

Cervical vertebrae: There are 7 cervical neck vertebrae. The uppermost of them is known as the atlas whereas the second is known as the axis. The other five are almost identical to each other. The cervical vertebrae differ from the other vertebrae in that they contain a pair of transverse foramina to carry arteries to the brain. The atlas supports the head and is attached to the occipital condyle. This pivot joint enables limited movement of the head in all directions. The axis is the second vertebra and enables axial movement of the head.

Disks

Thoracic vertebrae: There are 12 thoracic vertebrae. Each vertebra has three costal facets on each side that provide attachment for the ribs.

Lumbar vertebrae: There are 5 lumbar vertebrae. They are the largest and strongest vertebrae since they carry the weight of the body. The articular process of the fifth lumbar vertebra differs from the others in that the joint between it and the first sacral vertebra is more flexible than the joints between the other lumbar vertebrae.

Sacral vertebrae: There are 5 fused sacral vertebrae during the first year of development. After this period they fuse together to form a triangular structure known as the sacrum. This structure is joined to the fifth lumbar vertebra on its upper surface and laterally to the ilium of the pelvic girdle. There are four pairs of sacral foramina on the dorsal convex surface of the sacrum. The spinal nerves protrude from these foramina.

Sacral foramina

Figure: Lateral and dorsal view of vertebral column which is composed of 33 vertebrae.

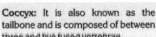

Coccyx: It is also known as the tailbone and is composed of between three and five fused vertebrae.

c. Thorax

It is a part of the axial skeleton and is also known as the chest. It functions in the protection of the heart, lungs and other abdominal organs. It consists of 12 pairs of ribs and a sternum. The ribs are connected to the thoracic vertebrae on their costal faces. Although all are directly connected to the vertebrae, in the abdomen the first

7 pairs are joined directly to the sternum (true ribs) whereas the 8th, 9th and 10th ribs are joined via the 7th ribs (false ribs). The 11th and 12th ribs are free--they are not connected to the strenum (floating ribs).

The sternum is the midline bony structure located between the left and right ribs in the center of the anterior chest wall.

1. Pectoral girdle

Each side of the pectoral girdle consists of a clavicle which is joined to the sternum and a scapula.

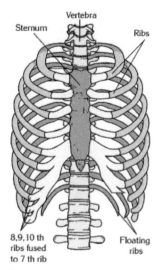

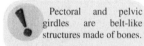

Figure: Vertebral column and rib cage.

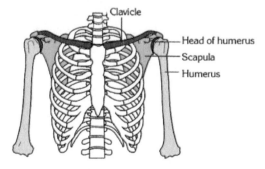

Figure: Transverse section through the trunk, illustrating the skeletal structures.

> **!** Pectoral and pelvic girdles are belt-like structures made of bones.

2. Pelvic girdle

It consists of the ilium, ischium, pubis, sacrum and coccyx.

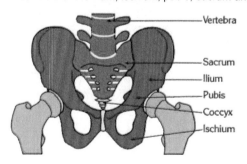

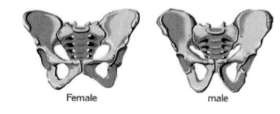

Figure: The structures that form the pelvic girdle.

UPPER EXTREMITIES

Arm bones		Hand bones	
Humerus	(1)	Carpals	(8)
Radius	(1)	Metacarpals	(5)
Ulna	(1)	Phalanges	(14)

Arm bones

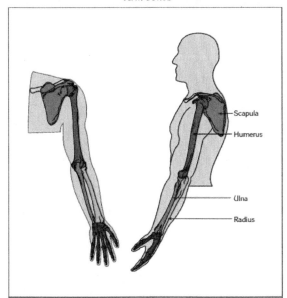

Hand bones

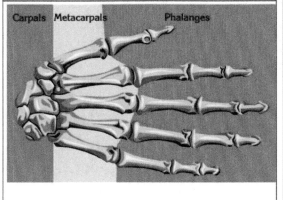

UPPER EXTREMITIES

Leg bones		Foot bones	
Femur	(1)	Tarsal	(7)
Tibia	(1)	Metatarsals	(5)
Fibula	(1)	Phalanges	(14)

Leg bones

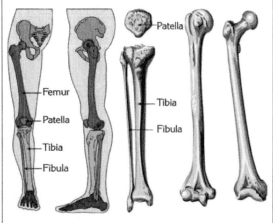

Foot bones

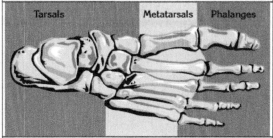

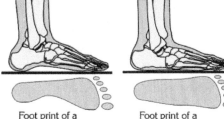

Foot print of a normal individual

Foot print of a Flat-footed individual

d. Appendages

1. Upper extremities

They consist of two arms and the substructures that join them to the body at the pectoral girdle. The bones of the upper extremities are 30 in number, and comprise the following;

Humerus

It is a long bone joined to the scapula on its upper side and joined to the radius and ulna on its lower side.

Hand bones

The hand is composed of carpals, metacarpals and phalanges. The eight bones of the carpals are located in two orders. The upper order is joined to the radius of the forearm and the lower order is joined to the metacarpals. The five metacarpals are joined to the phalanges. All phalanges except the thumb are composed of three bones.

2. Lower extremities

The two legs and feet, which form the lower extremities, are composed of 30 bones and are attached to the pelvis by the femur.

Femur: It is a strong bone connecting the knee and the hip and is the longest bone of the body. It is our strongest bone. It withstands pressures of 540 kilograms per 2.5 cm^3 when we walk.

The pyramidal bone known as the patella forms the knee cap and is joined to the femur.

Shin: It is between the knee and ankle and includes the tibia and fibula.

Foot: It consists of tarsal bones, metatarsal and phalanges. The tarsals on each foot are formed from 7 bones.

5. Joints

Joints are connection points of two or more bones. There are three types of joints.

Immovable

Slightly movable

Movable

a. Immovable Joints

They contain structures or interconnections which hold the closely positioned bone plates together.

All cranial and facial bones with the exception of the mandible are immovable.

As the strongest bone of the body, the femur withstands

pressures of 540 kilograms / 2.5 cm^3

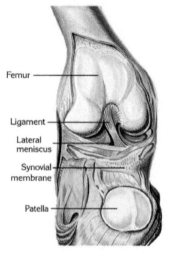

Figure: Knee joint is an example of moveable joints.

Femur

Ligament

Lateral meniscus

Synovial membrane

Patella

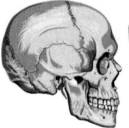 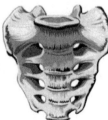

Figure: Cranial bones and lower part of vertebral column are immovable.

Locomotion Systems

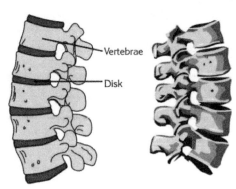

Figure: Cartilage disks between vertebrae let them move slightly.

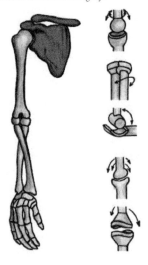

b. Slightly movable joints

Each component of the slightly movable joint is separated by either cartilage or connective tissue. The degree of movement is governed by the elasticity of the interconnecting material. Examples include cervical, thoracic and lumbar vertebrae.

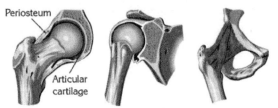

c. Movable joints

Movable joints include most joints of the body and are responsible for movement. They are generally synovial joints and can be subdivided into the following:

Hinge: this forms the junction of two bones. This type of joint allows movement about one axis. The elbow is an example of this type of joint.

Cylindrical: this also forms the joint of two bones. More movement is permitted in this type of joint. The junction of the atlas vertebra with the occipital bone is an example of this type of joint.

Ball and socket: This type of joint allows extensive movement, such as rotation in many directions. Examples include the joint of the shoulder and hip.

The structure of a movable joint:

A hinge joint is the junction of two bones allowing movement in one plane. The point of articulation of each bone is covered by articular cartilage composed of hyaline cartilage. The synovial membrane forms a capsule surrounding the joint and contains synovial fluid which cushions, providing a friction-free environment. The joint itself is held together by strong ligaments.

READ ME

Scurvy

Infantile scurvy is a deficiency disease due to the lack of water-soluble vitamin C. It may be found in both children and adults. The infantile form has a more acute onset and the character of the bony changes is influenced by the undeveloped state of the affected bones. The essential pathogenic feature is the inability of mesenchymal derivatives to form collagen-like cement substances in the absence of ascorbic acid. Thus the osteoid substances are not elaborated, and the cement substance of endothelium is lacking. The skeletal structures revert to a primitive fibroblastic pattern and the vessel lining is extraordinarily fragile and permeable. Persisting spicules of bone are of a delicate, fragmentary variety. Infantile scurvy begins most frequently between the sixth and tenth months of life and reaches an acute stage in two or three months. In children thus affected, motion is extremely painful and there is a pseudoparalysis because of the immobile state in which the limbs are held. Hemorrhages in the neighborhood of the bone are an outstanding characteristic of the disease. X-rays of the affected bones depict a marked increase in the density of calcification, or "white line," behind the epiphyseal line, and the epiphysis may have a characteristic ring about its edge. Behind the dense line, areas of rarefaction are frequently seen, and at these sites epiphyseal separations may take place. Subperiosteal hemorrhages followed by a slight amount of new bone formation are seen in advanced cases.

What is Osteopathy?

Osteopathy is the modern, scientific development of two of the oldest forms of treatment known to man--massage and manipulation. It is essentially a natural therapy, which seeks to overcome the wide range of disease, disabilities and pains which result from disturbances of the body's framework and moving parts.

How does Osteopathy work?

Just as structural engineers undergo a lengthy training to help them understand the mechanics of bridges, dams and high rise buildings, so osteopaths follow an extensive, four year training including anatomy, physiology and pathology of the human body. This equips them to analyze your problems and diagnose your complaints, using a variety of clinical skills backed up where necessary by X-ray examinations and biochemical tests. Their treatment, designed to correct the faults revealed by this thorough structural investigation, is gentle and rarely causes pain. In most cases it is followed by explanation and advice to help you prevent a recurrence of your trouble. If other treatment is indicated, you will find that your osteopath will refer you to the most appropriate source of help.

Choosing your Osteopath.

In many cases today you will be referred to an osteopath by your general practitioner. At other times you may be recommended by a friend who has reason to be grateful for an osteopath's help.

Where can an Osteopath help?

Osteopathy is helpful in overcoming stiffness and locomotor disabilities (problems with walking, climbing stairs, dressing and getting up from beds and chairs). But most of all it is effective in alleviating acute and chronic pain.

Athletic injuries

The growth of interest in sports and fitness training has led to an increase in sports-related injuries, which osteopaths are uniquely equipped to handle. Many of the world's top athletes today acknowledge the debt they owe to osteopathic treatment.

Muscular rheumatism

Muscle pains, lumbago, "fibrositis" and rheumatic aches in the legs and arms, are some of the most common medical complaints. These conditions can be alleviated, but rarely cured, by drugs. To obtain more permanent relief it is necessary to try and eliminate their underlying cause, something the osteopath is specifically trained to do.

Foot pain

The foot is an incredibly efficient piece of human engineering; a complex arrangement of twenty-six bones united by ligaments to form thirty-three separate joints. It is also a frequent source of pain, which can be severe enough to make walking difficult and life miserable, a disability which can often be relieved by skilled osteopatic treatment.

Tension headeches

It is generally recognized that a large proportion of headaches - 80 per cent or more - originate from stiffness and tension in the neck. This gives rise to pain, often radiating over the skull from the base of the neck to the eyes. Heat and analgesic tablets frequently ease this pain, but long-term relief often requires accurate osteopathic diagnosis and treatment of the underlying problem.

Back pain

Surveys reveal that four out of five Britons will be crippled by backache at some time in their lives. This is one of man's perennial problems, a malady as widespread as the common cold, as painful as toothache and sometimes as crippling as a stroke. The vast majority of back pains result from mechanical disturbances of the spine - postural strains, joint derangements, spinal disc injuries - and can be relieved or completely cured by carefully designed osteopathic treatment.

Nobody should allow themselves to become a martyr to chronic back pain without consulting an osteopath. More and more people are taking this step and being delighted by the results. (A MORI poll revealed that 63% of patients visiting osteopaths were "very satisfied" with the treatment they received.)

Approximately a third of Britons suffer an episode of backache in any given fortnight. This is a major cause of sickness absenteeism. Many workers spend weeks in bed before they finally recover from their back pain; others find that a visit to an osteopath can dramatically shorten their recovery time.

If you are suffering pain in your back, neck, head, legs or arms, seek the help of a registered osteopath who is trained to diagnose your trouble and provide the treatment you need.

Locomotion Systems

Bones are extremely strong. However, if exposed to a sudden impact or violent movement, they can crack or break. The older the bone, the more likely it is to break under impact. A young bone is more likely to bend or twist under pressure - a condition known as distortion. Fractures may be either closed or open. In both cases the casualty is incapable of moving the effected area, which is tender and swollen and may appear deformed. Any movement of the body close to the area results in considerable pain. A closed fracture of the arm should be immobilized and supported by a splint, then bound to the body in an elevated position using a sling. An open fracture should be covered by a sterile dressing and elevated to reduce any bleeding. More serous bleeding can be slowed by applying pressure to the wound. As for a closed fracture, the limb should be immobilized.

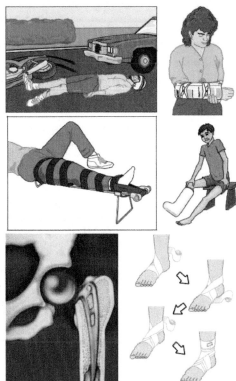

Dislocation

Dislocation results when an intact bone is moved from its original position by strong force such as through a fall or by a sudden muscular contraction. The most commonly affected bones of the body are the fingers, shoulders, wrists and jaw. Symptoms of a dislocation are severe pain in the region of the affected joint, an inability to move the affected part and visible deformation at the site of injury.

Treatment

The dislocated area should not be moved.

The area should be immobilized and supported using cardboard or other stiff material and a bandage.

A dislocated shoulder should be immobilized with a sling.

The casulty should be taken to hospital as quickly as possible.

WARNING: Never try to move either the joint or the patient to understand if the injury is due to a dislocation or a fracture. Never attempt to move the head or back of the casualty if there is any suspected injury to the spine. During movement, any piece of bone may damage or sever the spinal cord, resulting in paralysis.

Keep the person who has neck or vertebral column dislocation in the same position you meet and don't move the patient.

Call the doctor as soon as possible.

Distortion

If a joint is forced to carry excessive weight, damage can result to the surrounding tissue. This is known as distortion. As the level of damage to the joint increases, ligaments, tendons, muscles and blood vessels are compressed or torn. The joints of the fingers, toes, wrists, ankles and knees are most affected.

Symptoms:

Inflammation around the affected joint   Sensitivity   Bruising

Note: Fractures and distortions generally have similar symptoms and can only be distinguished between by X-ray film.

Treatment

Cover the affected area with a cold compress.

Immobilize the joint.

Call for medical attention or take the casualty to hospital.

WARNING: If the vertebral column is affected, do not attempt to move the patient until medical help arrives.

HUMAN BIOLOGY

98

MUSCLES

Most animals move by the contraction of specialized muscle cells. However, some unicellular organisms such as ameoba, and single-celled structures such as leucocytes, move by amoeboid motion. In this type of movement, the cytoplasm flows out to form projections, or pseudopodia. The rest of the cytoplasm also flows to this side, resulting in motion. Most bacteria, sporozoa and sperm cells move using flagella.

In multicellular organisms, there is cooperation and coordination between cells. Locomotion, for example, is provided by specialized muscle tissue. It is composed of specialized cells, known as **myofibrils,** which have an ability to contract and relax. Muscle tissue is divided into three groups, smooth, striated skeletal and striated cardiac according to its morphological and physical structure as well as its location in the body.

1. Muscle Tissue

a. Smooth muscle

Each cell is long, fibrous and sharp-ended with a single central nucleus. Cells form cords which project in different directions. Smooth muscles are generally regulated by the autonomic nervous system (ANS) since they are located in organs that function involuntarily. Their function is generally irregular and slow.

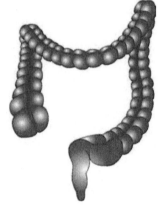

A smooth muscle cell consists of a membrane known as the **sarcolemma** and cytoplasm known as **sarcoplasm**. The sarcoplasm contains unstriated myofibrils. Smooth muscle cells participate in the structure of internal organs of vertebrates and the whole body structure of invertebrates.

Figure: Intestine has smooth muscles in its structure.

b. Striated skeletal muscle

They are involved in the skeletal system. They cover the skeleton. They provide locomotion in cooperation with the skeletal system. Cells of striated muscles are long, cylindrical and multinucleated.

They are termed muscle fibers due to their morphological appearance. Their fibers are mostly nonbranched. Each fiber of skeletal muscle is coated by a continuous layer of thin membrane known as sarcolemma.

The structures of striated skeletal muscles are listed in decreasing order of size as muscle bundles, muscle fibers, small fibers and actin and myosin proteins (myofilaments).

The fibers composed of actin and myosin proteins are arranged in parallel lines forming dark and light bands. This type of muscle is named for its striated appearance.

Striated skeletal muscle is voluntary and functions under the control of the brain. It contracts rapidly as compared to smooth muscle. Striated skeletal muscle is involved in the physical adaptation of the organism to the environment by movement.

Figure: Striated muscles cover the skeleton.

Muscles of the forearm: The flat, thin and long muscles of the forearm the involved in the movement of the hand and finger. Additionally, the muscles running in the same plane from the elbow to the wrist assist in rotation of the forearm.

Muscles that move the hand: They are small, short muscles which occupy the spaces between phalanges. They are involved in the movement of fingers.

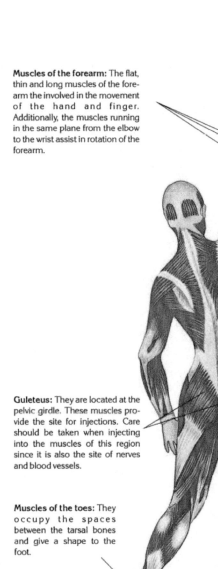

Guleteus: They are located at the pelvic girdle. These muscles provide the site for injections. Care should be taken when injecting into the muscles of this region since it is also the site of nerves and blood vessels.

Muscles that move the leg: The anterior region of the femur is surrounded by quadriceps femoris muscles. They are involved in the extension of the leg at the knee joint. Furthermore, the biceps femoris, semitendinous and semimembranous muscles flex and rotate the knee joint.

Muscles of the toes: They occupy the spaces between the tarsal bones and give a shape to the foot.

Muscles that move the foot: They are located in the anterior region of the foot and are involved in the movement of the foot and toes. The gastrocnemic is the strongest muscle of the foot which functions in flexing of the foot.

Muscles of mastication: The muscles involved in the chewing are the masseter temporalis, lateral pterygoid and medial pterygoid. During grinding motion, the mandible may move axially by the activities of masseter and temporalis in the same direction. The lateral and medial pterygoid muscles move in the opposite direction.

Muscles that move the tongue: They are composed of the genioglosus, styloglosus, hyoglosus and intrihsia muscles. The genioglosus protrudes and depresses the tongue However, the hyoplosus elevates the tongue. The intrinsic muscles alter the shape of the tongue.

Muscles of the abdominal wall: The muscles of the anterior posterior and lateral wall are involved in bending forward and backward. They are also involved in twisting of the back.

Muscles of the scalp: They are involved in the facial expression of individual.

Muscles of the neck: The muscles of neck function in neck movement and its support.

Muscles that move the humerus: The main muscles of humerus are the triceps and biceps. They cooperate in the movement of the humerus.

Muscles that move the shoulder girdles: They are involved in the movement of the arm and ribcage. Intercostal muscles relax expanding the chest cavity during respiration.

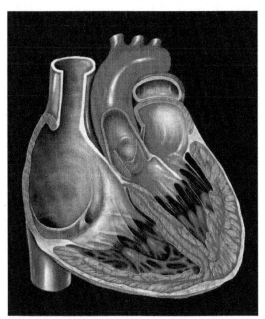

Figure: Myocardium of the heart is composed of cardiac muscles.

c. Cardiac muscle

Although it is structurally similar to striated muscle, there are some differences.

The nucleus is located at the center of each cell.

More mitochondria are present than in striated skeletal muscle.

Fewer myofibrils are present in the sacroplasma.

Each cell has branch-like projections

Each cell is rich in blood and lymph vessels.

The cardiac muscles perform their functions involuntarily under the control of the autonomic nervous system. Cardiac muscle has no capacity to divide or regenerate since it is composed of highly specialized cells. If damage occurs to the muscle, it is substituted by connective tissue.

The properties of muscular systems:

The stimulus is transmitted along the muscle fibers.

The response to the stimulus is muscle contraction and relaxation.

Chemical energy is converted to mechanical energy during the response.

Locomotion and support is provided.

2. The Muscular System of Vertebrates

Vertebrates have both smooth and striated muscle in their body structures. Striated muscle articulates with the skeletal system to provide movement, whereas smooth muscle is involved in the activities of internal organs.

a. Skeletal Muscles

They constitute 40% of the body. Striated skeletal muscles are composed of even more highly specialized cells than the nervous system. The striated muscle tissue is composed of vertically striated fibers and connective tissue which occupies the internal space.

The tips of striated muscles are coated by strong connective tissues termed **tendons**. There may be, however, more than one tendon at the tip of a muscle. Tendons are unique to skeletal muscles and connect them to the bones. They are more resistant as compared to the muscles however, and have no ability to contract. Each skeletal muscle fiber is composed of many muscle cells. These muscle cells elongate through the muscles. The simultaneous contraction of all cells produces muscle contraction

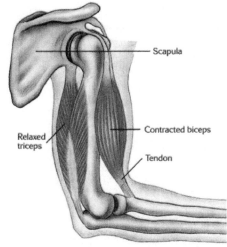

Scapula

Relaxed triceps

Contracted biceps

Tendon

Figure: Movement is the result of two sets of muscles acting antagonistically. While the biceps muscle contracts the triceps relaxes.

The surface of skeletal muscle is surrounded by a tight membrane of connective tissue which protects the muscle fibers and adheres them to each other.

Skeletal muscles provide movement by articulating with a joint. Joint-dependent muscles are grouped according to their direction of movement. Muscles that act cooperatively in the same direction are known as synergic, while those that act in the opposite direction to each other are termed antagonistic. For instance, we use a group of muscles when we close or mouth (synergy). On the other hand, when we bend the arm we use our biceps muscle; the triceps muscle returns it to its original position. Thus, the bicep and the triceps are antagonistic. Muscles that bring bones nearer to each other are called **flexors** (e.g. biceps), and the muscles that move bones apart from each other are called **extensors** (e.g. triceps).

Muscles at rest are under tension and are said to have **tonus**. This slightly contracted state provides a permanent state of readiness for contraction. The muscles of the body are never, therefore, in a completely relaxed state. Muscle tonus is completely destroyed if the motor neuron at the muscle fiber is fatigued.

b. The Muscular Contraction

The muscle is stimulated by impulses from the brain or spinal cord. The point of attachment of muscle and a motor neuron is known as **neuromuscular junction**. The point at which the nerve ending is attached to the muscle membrane is termed the **motor end plate**. When the impulse reaches the motor end plate, it produces a potential difference in this region, thus stimulating muscle fibers. If the flow of impulses from the neuron to the muscle is inhibited, paralysis results, as the muscles are not stimulated. The muscles can only contract in the presence of a stimulus.

No contraction or movement is observed in a muscle if it is excited by only a weak impulse. If the potential is gradually increased to the threshold level, an impulse is generated when this level is reached. Any excitement beyond this value has the same effect on muscle contraction. The energy for contraction is supplied by the muscle cells, not by the action potential. The response of the muscle at a certain threshold level is known as the all-or-none rule.

Muscular contraction occurs in three phases

Latent phase is the interval between the stimulation of muscle and the initiation of contraction. It lasts approximately 0.005 seconds.

Contraction is the interval between the initiation of contraction and the initiation of relaxation. It lasts approximately 0.04 seconds.

Relaxation is the interval between the initiation of relaxation and restoration of the original position.

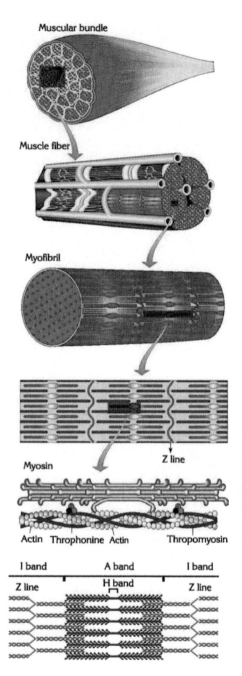

Muscular bundle

Muscle fiber

Myofibril

Myosin

Z line

Actin   Throphonine   Actin   Thropomyosin

I band	A band	I band
Z line	H band	Z line

Locomotion Systems

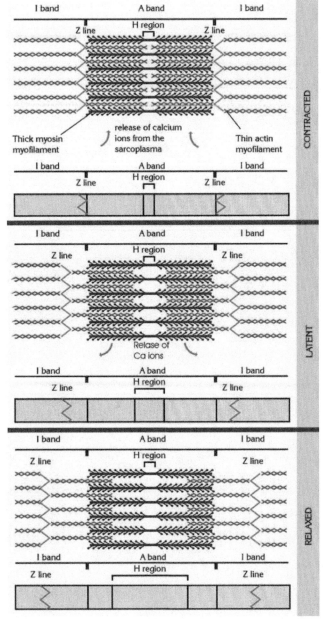

I band | A band | I band
Z line | H region | Z line

Thick myosin myofilament

release of calcium ions from the sarcoplasma

Thin actin myofilament

CONTRACTED

I band | A band | I band
Z line | H region | Z line

I band | A band | I band
Z line | H region | Z line

Relese of Ca ions

LATENT

I band | A band | I band
Z line | H region | Z line

I band | A band | I band
Z line | H region | Z line

I band | A band | I band
Z line | H region | Z line

RELAXED

Figure: During contraction A band is not changed. H band disappears. I band shortens.

c. The Mechanism of Contraction

The muscle contracts by actin and myosin fibers sliding over each other. Myosin is a protein which is 100–150 Å in width and 1.5 microns in length, whereas actin is 50–70 Å in width and 2 microns in length. Thus, myosin is shorter but thicker than actin. The muscle is in a relaxed state when the actin and myosin fibers move apart. The tips of each myosin fiber are thick with protruding branches (head of myosin). They give a folded appearance to each fiber.

Actin fibers consist of three types of protein. The most abundant is the spherical actin protein. Approximately 300-400 are present in each fiber. There are also 40-60 tropomyosin molecules in addition to a small amount of circular tropin molecules.

A muscle has repeating units called sarcomeres, which extend between two dark lines called the Z lines. The thick filaments are made up of a protein called myosin, and the thin filaments are made up of protein called actin. The I band is light colored because it contains only actin filaments attached to a Z line. The dark regions of the A band contain overlapping actin and myosin filaments, and its H zone has only myosin filaments.

The plasma membrane of a muscle fiber is called sarcolemma, the cytoplasm is the sarcoplasm and the endoplasmic reticulum is the sarcoplasmic reticulum. A muscle fiber also has some unique structures, for example T tubules (transver tubules), which are extensions of sarcolemma that penetrate into the cell so that they come into contact with special portions of the sarcoplasmic reticulum, called calcium storage sacs, containing calcium ions (Ca^{++}).

Chemical changes during contraction

Energy from ATP is required for the contraction of muscles. When the impulse reaches the sarcolemma, it is transmitted to the inner portions of muscles (to the calcium storage sacs) via the T tubules. Ca++ ions in the sarcoplasmic reticulum flow into the sarcoplasm, increasing its concentration. Calcium ions released from the storage sac combine with troponin.

After binding occurs, the tropomyosin threads, which wind about an actin filament, shift their position, and the myosin binding side is exposed. By using ATP molecules, cross-bridges between actin and myosin are formed. The thin actin molecules are pulled to the center of the sarcomere. This is contraction. When another ATP molecule binds to a myosin head, the cross bridge is broken, as the head detaches from actin (relaxation). During the contraction process, the I bands shorten and the H zones almost or completely disappear

If the muscle is stimulated again, after the refractory period (latent phase), but before it is fully relaxed, the second contraction adds to the remains of the first one, increasing the contraction strength. If the muscle receives repeated strong stimulations without a time to relax, the muscle has a continous contraction called a tetanus. Cramps, for instance, are a type of tetanus occurring in human muscular tissue and result from continuous stimulation of the muscles. This condition is extremely painful and potentially dangerous if an attack occurs while an individual is swimming, for example. Tetanus ends when the muscle uses up its chemical energy reserves. One of the symptoms of tetanus caused by the bacterium *CLOSTRIDIUM TETANI* is prolonged contraction of the muscles.

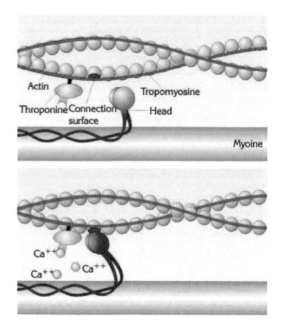

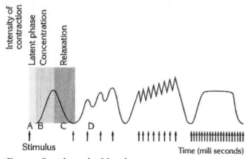

Energy Supply to the Muscles

The energy reserves in muscle can only supply energy for 0.5 seconds. After this period, creatine phosphate is used as an energy source. It is found in muscle cells and supplies 20 times more ATP energy. At rest, the muscle cells store the excess energy (ATP) that is unused, in the form of creatine phosphate. When it is needed, creatine phosphate donates its phosphate to ADP to form ATP, and ATP is used for muscular activity. When both ATP and creatine phosphate are consumed, glycogen and fatty acids stored in muscle cells are used as fuel for muscular

Figure: The steps of the chemical reaction that results in muscle function.

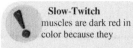

Slow-Twitch muscles are dark red in color because they contain myoglobin.

Fast-Twitch fibers are very strong, and white in color since there is no myoglobin in their structures.

activity. Use of food as a source of energy can be done by using oxygen (aerobic respiration), which produces much more energy (38 ATP / glucose molecule), or by using only enzymes but not oxygen (anaerobic respiration or fermentation), at the end of which only 2 ATP net gain can be achieved for each glucose molecule.

Slow-Twitch and Fast-Twitch Muscle Fibers

Muscles provide energy by means of both aerobic and anaerobic respiration. Some fibers, however, utilize one method more than the other. Slow-twitch fibers tend to be aerobic and fast-twitch fibers tend to be anaerobic.

Slow-twitch fibers are most helpfull in sports like long-distance running, biking and swimming. Because they produce most of their energy aerobically, they tire only when their fuel supply is finished. These muscles have many mitochondria and are dark in color because they contain myoglobin, which store oxygen, like hemoglobin molecules in blood. Fast-twitch fibers seem to be designed for strength because their motor units contain many fibers. The end of the motor neuron connects to many fibers. As a result, many fibers can contract at once, which makes the muscles very strong. They provide explosions of energy and are most helpful in sports like sprinting, weightlifting, etc.

These muscles are white in color because they have fewer mitochondria, little or no myoglobin and fewer blood vessels than slow-twitch muscles. Lactic acid, which is a by-product of anaerobic respiration, accumulates between the muscle fibers and causes them to fatigue quickly.

READ ME *ANTA*

Muscles and bones work in coordination to provide movement. During movement, bones act as levers and give mechanical support to movement. Although muscles move within a restricted area, they transport their movement to the joints. For example, biceps muscles contract 80-90 mm, but provide movement of 50-60 cm.

Muscles can only contract and relax. When contracted muscle relaxes, it must return to its original position. This explains why muscles move in reverse to each other. When a muscle moves to one side, the other one moves to the opposite side. When a muscle is contracted, the other is relaxed. This type of muscles are called antagonistic muscles (for example biceps, triceps). When the biceps contract, the triceps relax.

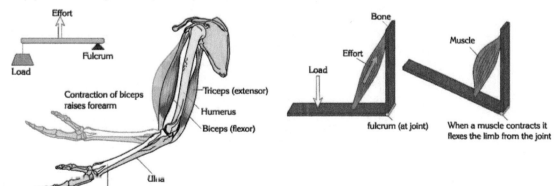

HUMAN BIOLOGY

106

Exercise and Size of Muscles

Forceful muscular activity over a prolonged period causes muscle to increase in size. This is called hypertropy and it occurs only if the muscle contracts to at least 75% of its maximum tension. Some people takes drugs, such as anabolic steroids, testesterone or related chemicals to promote muscle growth. This practice, however, can cause cardiovascular diseases, liver and kidney malfunctions and sterility.

Rigor Mortis

Any metabolic disorder in a muscle results in the loss of Ca^{++} ions from the sarcoplasmic reticulum. This process is irreversible, the muscles remaining contracted. This fact explains, why after death, all the muscles of the body contract.

Rigor mortis first occurs in the heart, diaphragm, face, neck, jaw and eyelids which, in humans, do not close after death. The sequential hardening and relaxation of muscles during rigor mortis may be used as evidence in forensic medicine when a death is thought to be suspicious. If, however, a muscle is removed and placed into an oxygen-rich environment immediately after death, rigor mortis does not occur. The muscle fibers are autolyzed by catepsin after the process of rigor mortis is complete.

 Medicines such as anabolic steroids can promote muscle growth.

However, using such medicines can cause very dangerous diseases, such as liver and kidney problems.

READ ME

Inflammatory Diseases of Muscle

Infectious and Traumatic Myositis

Inflammatory diseases of the muscle may result from infection or trauma. In most instances suppurative myositis is secondary to a systemic disease, such as a generalized septicemia, which produces focal metastatic abscesses, or it occurs from extension from a suppurative process in the neighboring bones, joints and periosteum. Inflammatory and degenerative changes are found in the muscle in typhoid fever and in leptospirosis (Weil's disease). In pyogenic osteomyelitis, in tuberculosis of the bones and joints, and in syphilitic periostitis and oseitis, the infectious process often extends to the adjacent muscles. This is particularly true when bony sequestra are discharged. When pyogenic or tuberculous osteitis involves the vertebrae, abscesses may form along the fascial planes between the muscle bundles (retropharyngeal and psoas abscesses are characteristic examples). Focal infiltration of leukocytes and fibrosis of the fascial sheath is accompanied by degenerative changes in the adjacent muscle. Fragmentation of the fibers and swelling of the muscular nuclei gives rise to a characteristic pseudo-giant cell.

The most common form of traumatic myositis results from tearing of the muscle fibers with strenuous activity. Hemorrhage, edema and fibrosis occur, but only temporary pain and disability are experienced by the patient. Such acute traumatic myositis is commonly known as a "charley horse." With faulty posture, scoliosis and repeated trauma or heavy lifting, a chronic myositis with fibrositis may result. This is described above under the heading of "myofascitis."

Myositis Ossificans

This is a rapid ossification of muscle tissue following one to two months after trauma or occasionally appearing as a hereditary congenital abnormality.

Thorium Dioxide Myositis

When thorium dioxide is used as a contrast media for X-ray diagnosis, spilling of the material into the muscles may occur. This causes characteristic changes. There is extensive fibrosis and a characteristic deposition of brown amorphous material. Alpha particles, a form of radiation highly damaging to the tissues, are emitted by the conversion of thorium to mesothorium.

Dermatomyositis

This is a rare disease characterised by edema, skin eruption resembling toxic erythema, and progresive involvement of groups of muscles. Rarely, similar changes may be a complication of disseminated lupus erythematosus. It is usually seen in childhood and is often fatal.

Locomotion System

LOCOMOTION SYSTEMS

A. Key Terms

Axial skeleton	Spongy bone
Compact bone	Osteocyte
Epiphysial plate	Red bone marrow
Actin	Myofibril
Myoglobin	Myosin
Sarcomere	Tendon

B. Review Questions

1. Explain the effects of an external skeleton on growth, and compare a shell with skin.

2. Compare an exoskeleton and an endoskeleton.

3.

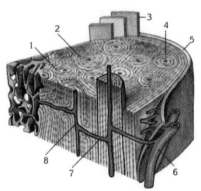

Label the components of bone tissue in the figure.

4. Explain the function of the Haversian canal and the Volkman canal.

5. Which component of bone is involved in longitudinal growth?

6. Compare red bone marrow and yellow bone marrow.

7. List the factors involved in bone formation.

8.

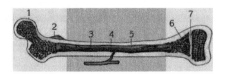

Label the components of bones as shown in the figure.

9. What is the origin of red blood cells in long bones?

C. True or False

1. Creatine phosphate is a compound found in muscles and responsible for energy supply.

2. The contractile portion of a muscle fiber is called myofibril.

3. Fibrous connective tissue that joins muscles to bone is ligament.

4. Hard bone is also known as spongy bone.

5. The connection sites of bones are known as joints.

D. Matching

a. Osteocyte ( ) Portion of the skeleton forming the girdles and appendages.

b. Ligament ( ) Mature bone cell.

c. Appendicular skeleton ( ) Fibrous connective tissue that joins bone to bone at a joint.

d. Sarcolemma ( ) Steady, prolonged muscle contraction

e. Tetanus ( ) Cell membrane of a muscle fiber.

Brain controls overall activities of muscles. Muscles cover and protects the nerves.

Skeletal System

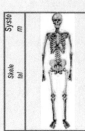

Muscles contract and relax to provide movement of skeleton. Muscles also cover and protect the skeleton. Bones provide attachment sides for muscles and store Ca^{++} needed for muscles to contract.

Muscles protect glands. Testesterone promote skeletal muscle growth. Epirephrine (Adrenalin) stimulates hearth and blood vessel constriction.

Smooth muscle contraction control urination and movement of urine. Skeletal muscles protect excretory organs.

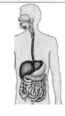

Peristaltic movement is provided by smooth muscles.

Skeletal muscles protect organs of digestive system.

Contraction of ribmuscles and diaphragm assist breathing. Lungs provides O_2 to and carry away CO_2 from muscles.

Muscle contraction provides blood movement within the heart and blood vessels. Blood delivers nutrients and oxygen to the muscles and remove wastes from them.

Bones protect the brain and spinal cord. Storing Ca^{++}, which is needed for transmission of impulses, is also provided by the skeleton.

Movement of the body is provided by the skeleton, this is controlled by nervous system.

Skeletal system provide protection and support. Kidneys aid regulation of mineral concentration.

Muscles attach themselves to bones.

Bones store Ca^{++}, which is needed for muscular contraction. Muscles cause skeleton to move. Bones are covered by muscles, and are protected.

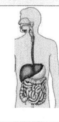

Bones provide support and protection for organs of digestive system.

Digestive system provide nutrients including minerals for skeleton.

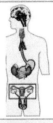

Bones protects the glands and store minerals which may be used in hormonal activities. Growth hormone, facilitates bone growth. Storing of Ca^{++} is done by means of Calcitonin, a thyroid gland secretion.

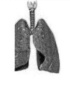

Rib cage protects lungs. Bones are sites of attachment for rib muscles, which are used for breathing.

As a result of respiratory process O_2 is supplied to the cells and CO_2 is removed.

Heart is protected by the rib cage. Red bone marrow produces blood cells.

Nutrients and O_2 are supplied to the skeleton by the blood, and metabolic wastes removed from the cells of skeletal system.

Circulatory System

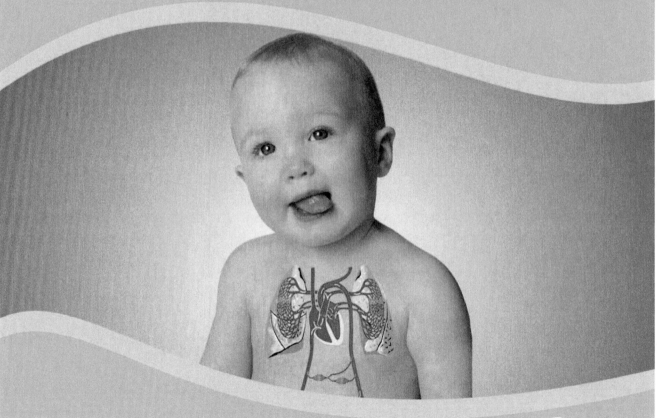

chapter **6**

CIRCULATORY SYSTEM

Organisms require transportation systems to bring in supplies (nutrients and oxygen for energy) and remove garbage (metabolic wastes) without disturbing their internal environments. Simple organisms can solve this problem without using any complex systems. By means of diffusion through their body surfaces, needed substances can be taken into their cells or bodies, and wastes can be removed by the same method. Animals more complex than flatworms don't have a great enough surface-to-body ratio to solve their problems. They have to use special, more complex sytems. The circulatory system in large active animals carries vital materials to cells and removes wastes.

A circulatory system includes three main structures: a pump, fluid and vessels. Circulatory systems transport fluid in one direction, powered by a pump that forces the fluid through vessels to all parts of the body.

Among animals, one of two types of circulatory systems can be used.

Open Circulatory Sytem (OCS)

In an OCS, the body fluid is not contained in vessels. Arthropods, gastropods and bivalves (types of mollusks) have open circulatory systems.

This system includes a heart, and blood vessels that lead to spaces where the fluid, hemolymph, directly bathes the cells. Material exchange occurs between the cells and fluid before the fluid returns to the heart. In such a system there are only two types of blood vessels: arteries, which carry fluid from the heart to the cells; and veins, which carry hemolymph back to the heart.

Closed Circulatory System (CCS)

In a CCS, blood remains within vessels. Arteries, leading from the heart, conduct blood from the heart and branch into smaller vessels, called arterioles, which then diverge into a network of very tiny, thin vessels called capillaries. Material exchange occurs between the capillaries and cells. Blood then collects into veins, which carry blood back to the heart. Annelids are the simplest animals with a closed circulatory system.

THE HUMAN CIRCULATORY SYSTEM

The human circulatory system is composed of the heart, arteries, capillaries and veins. All these structures are filled with blood, a fluid connective tissue composed of water, solutes, blood cells and platelets. Together they form an internal transport system within the body for substances to and from the cells. This system is also known as the cardiovascular system (cardio- means heart, while vascular means vessels).

1. Heart

The heart is located within the chest (thoracic cavity), between the lungs and under the sternum or breastbone. In adult males, the heart weighs approximately 280–340 grams, and in females, it weighs approximately 230–280 grams.

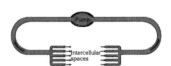

Figure: Open circulatory system. There are no capillaries between arteries and veins.

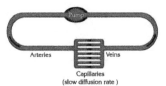

Figure: Closed circulatory system. There are capillaries between arteries and veins. Blood never leaves the blood vessels

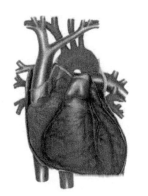

Each day the human heart sends 7000 liters of blood through the body, and it contracts more than 2.5 billion times in a lifetime. The heart is divided into two halves, forming the basis of the two cardiovascular pathways: the pulmonary circuit and the systemic circuit. Pulmonary circulation takes blood to the lungs and returns it to the heart. In pulmonary circulation, blood that is low in oxygen but high in carbon dioxide is pumped from the right side of the heart to the lungs where gas exchange takes place. The blood is oxygenated from inhaled air, and carbon dioxide diffuses out and is released by the lungs. From there the blood flows to the left side of the heart where the freshly oxygenated blood is distributed to the body by the systemic circulation. Oxygen is used by all parts of the body, and carbon dioxide is released as a waste product. The oxygen-depleted blood travels back to the right side of the heart where the process is repeated.

The heart is divided into left and right hemispheres separated by a muscular wall, the septum. Each half of the heart has two chambers: an atrium and a ventricle. The tricuspid, or three–flapped, valve connects the right atrium to the right ventricle and a bicuspid, or two–flapped, valve connects the left atrium to the left ventricle. Each half of the heart also has a valve known as the semilunar valve located between the ventricle and the arteries leading away from the heart. The function of all the valves is to prevent the backflow of blood and to keep the blood moving in one direction. The valves are unidirectional: they only allow blood flow into, and not out of, the ventricles. Any defect in these valves can result in heart malfunction.

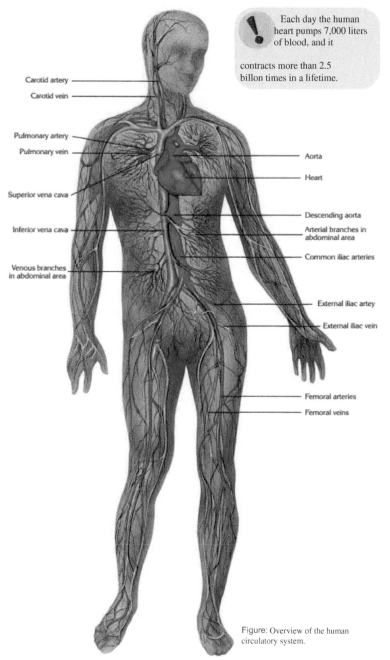

Each day the human heart pumps 7,000 liters of blood, and it contracts more than 2.5 billon times in a lifetime.

- Carotid artery
- Carotid vein
- Pulmonary artery
- Pulmonary vein
- Superior vena cava
- Inferior vena cava
- Venous branches in abdominal area
- Aorta
- Heart
- Descending aorta
- Arterial branches in abdominal area
- Common iliac arteries
- External iliac artey
- External iliac vein
- Femoral arteries
- Femoral veins

Figure: Overview of the human circulatory system.

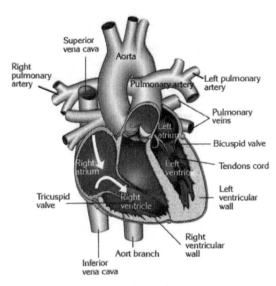

FIGURE: Components of a typical human heart.

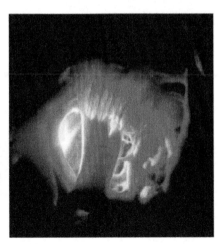

FIGURE: The heart strings attach the valves to the endocardium.

FIGURE: The mechanism of opening and closure. The tricuspid and bicuspid valves are located between the atria and ventricles. They are involved in the prevention of backflow of blood.

Oxygen-depleted blood is transported from the right ventricle to the lungs by the pulmonary artery. In the lungs the blood is oxygenated and returned to the heart by way of the pulmonary veins. The blood enters the left side of the heart and flows into the left ventricle where it is then pumped out to all regions of the body.

A. STRUCTURE OF THE HEART

The heart is composed of three main layers:

Endocardium

Myocardium

Pericardium

The endocardium, the innermost layer of the heart, is composed of a single layer of epithelial cells. It also contains connective tissue, connecting the endocardium to the myocardium. The endocardium contains no blood vessels.

Open valve

Additionally, its gelanitous structure prevents the erosion of the heart during contraction and relaxation.

The myocardium is the middle layer of the heart and is composed of cardiac muscle. It is the main layer of the heart, since the main function of the heart is that of a pump. The thickness of the myocardium varies. It is thin in the artia but thicker in the ventricles. The left ventricle however, has a thicker layer of myocardium than the right ventricle.

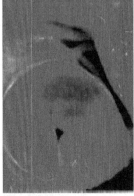

Closed valve

The cells of heart muscle do not obtain their nutrients from the blood within the heart chambers directly. The heart, as a hard-working organ, must be fed perfectly. Its nutrition is effected by a special branch of the systemic circulation, the cardiac circulation.

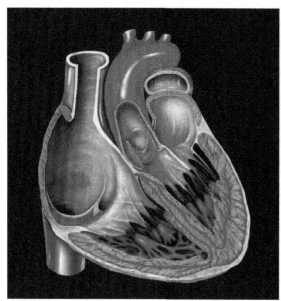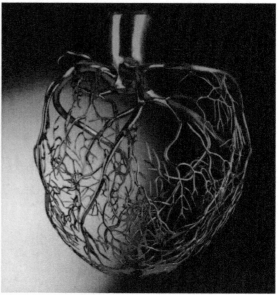

Figure: The heart is made up of three main layers: a pericardium, a Figure: Coronary blood vessels feed the cells of the myocardium. myocardium containing coronary vessels to feed the heart cells, and an endocardium consisting of a single layer of endothelial cells.

The Pericardium: This forms the outermost layer of the heart and is composed of fibrous tissue. The space between its two surfaces is filled with fluid. The colloidal structure of the pericardium facilitates heart function and protects it from external hazards.

b. The cardiac circulation

The heart cells are supplied by the circulatory system, consisting of coronary arteries, capillaries and veins. The aorta directs the oxygen- and nutrient-containing blood to the coronary arteries.

The blood is then distributed to the capillaries where nutrients and oxygen diffuse into the heart cells. Simultaneously, nitrogenous wastes and carbon dioxide diffuse into the blood. The deoxygenated blood, together with the waste, is collected by coronary veins which enter the right atrium.

Any disruption of cardiac circulation, such as a blood clot, may cause serious disorders. Blockage of the coronary artery in any case results in an infarction, in which blood is prevented from flowing.

c. Heart Nutrition

The oxygen requirement of the heart is greater than that of other tissues. Organisms need to have the ability to satisfy this requirement. New blood vessels are formed through exercise, resulting in higher blood flow in the coronary arteries and an increase in the amount of oxygen that is consigned to the heart.

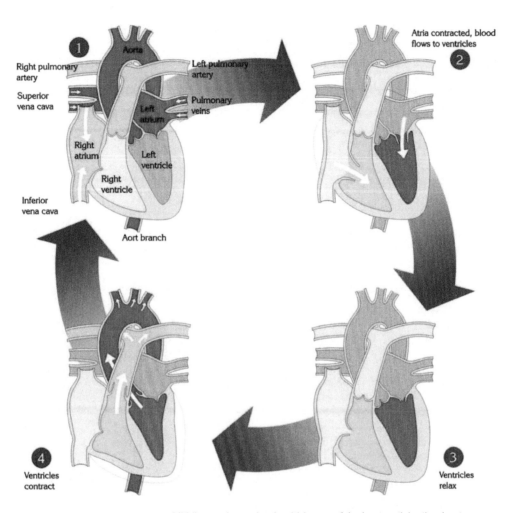

Figure: The cardiac cycle has two phases: systole and diastole. Rela-xation of the heart is known as diastole, while its contraction is termed systole. Blood is pumped into the ventricles through the valves by atrial contraction.

Middle-aged people should be careful when participating in strenuous sports or activities that require great physical exertion, since the formation of new blood vessels in the heart may cause serious disorders at that age. The heart consumes a greater amount of energy than other tissues. The main energy source of the heart comes from fatty acids. Approximately 67% of its energy needs are met in this way. During exercise, carbohydrates are used, since their usage is simpler.

d. Cardiac Activity

The heart functions as a pump for the circulation of blood within the body. It relaxes and contracts in order to send blood to the tissues. Relaxation of the heart is known as diastole, while its contraction is termed systole.

Blood is pumped into the ventricles through valves via atrial contraction. It is forced into the arteries by the subsequent contraction of the ventricles. The contraction of the atria and ventricles is reversed: the atria relax when the ventricles contract. Likewise, the ventricles relax when the atria contract.

Contraction of the right and left atria occurs at the same time. Both contract, pumping the blood into the ventricles. Back flow of blood is prevented by valves. The contraction of the ventricles is characterized by similar events. The ventricles contract and pump blood into the arteries through valves.

The specialized tissues of the heart that regulate its activities are as follows:

Sinoatrial node (SAN): This tissue consists of special muscle fibers located at the right atrium near the entrance of the superior vena cava. The other AV (atrioventricular) node is found in the base of the right atrrium very near the septum. The SA node initiates the heartbeat and automatically sends impulses every 0.85 seconds.

This causes the atria to contract. (This interval depends on the activity of the body. When an individual is exercising, the interval between each heartbeat decreases, leading the heart to beat faster.)

When the impulse reaches the AV node, the AV node signals the ventricles to contract by way of the bundle of fibers that branch and terminate in the more numerous and smaller Purkinje fibers.

The SAN is called the pacemaker because it initiates the heartbeat and usually keeps it regular. If it fails to work properly, the heart continues to work, but probably not regularly. In a such case, to correct the condition, it is possible to implant an artificial pacemaker which automatically gives electric stimulus to the heart every 0.85 seconds so that the heart beats about 70 times per minute.

The SAN, the AVN and the Purkinje fibers are special cardiac muscle cells which carry stimuli within the heart, just like nerve fibers. A nerve that attaches to the SAN initiates all the stimulus.

Then, all the other steps of carrying this stimulus to every part of the heart are done by this special system. With the contraction of any muscle, including the myocardium, ionic changes occur. These can be detected with electrical recording devices. The pattern that results is called an electrocardiogram (ECG or EKG).

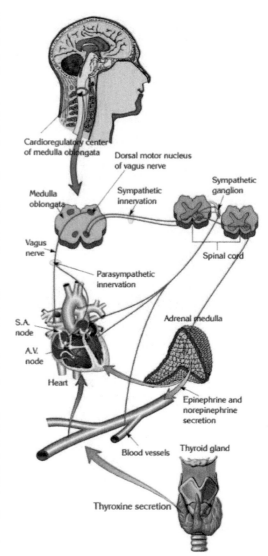

Figure: Showing the atria, ventricles, valves, fibrous cords, and coronary vessels. Under stressful conditions, heartbeat is increased by impulses generated by sympathetic nerves. Acetylcholine accelerates heart function and causes constriction of vessels to increase blood pressure. Furthermore, it stimulates the secretion of epinephrine and norepinephrine from the adrenal medulla. These hormones act by accelerating the heartbeat. Additionally, thyroxine secreted by the thyroid gland affects the heartbeat by increasing metabolic activity. Conversely, after stress or physical exertion, the vagus nerve generates impulses to reduce the rate of heartbeat.

Cardiac cycle is regulated by a special system within the heart.

This system includes: Sino Atrial Node (SAN), Atrio Ventricular Node (AVN) and Purkinje fibers. Messages coming from the brain stimulate these structures. The heartbeat rate and the harmony between systole and diastole are regulated.

e. Factors affecting heart function:

Nerves, hormones, temperature and severe diseases can all affect heart performance.

Capacity: Heart rate is directly related to the activities of the individual. A working organ consumes 7–8 times the quantity of food and oxygen as compared with a resting organ. As a result, the heart of a manual laborer should work 7–8 times more, since all the cells need raw materials to function.

Nerves: The SA node is stimulated by the sympathetic nerves of the autonomic nervous system, originating in the medulla. Parasympathetic nerves reduce the heart rate. The vagus nerve, a parasympathetic nerve, secretes acetylcholine, which slows down the pulse. Conversely, adrenaline, secreted from the sympathetic nerves, speeds up the heart rate

Hormones: The amount of adrenaline (epinephrine) in the blood increases in times of stress and excitement. Thus, the heart rate also increases. Furthermore, thyroxine released from the thyroid gland speeds up heart function.

Inflammatory diseases: These increase the heat of the body. Thus, the rate of heartbeat also increases.

READ ME

Infarction (Heart Attack)

Infarction is a common and serious disease. If insufficient nutrients and oxygen are supplied to the cardiac muscles, serious disorders in the affected region of the heart may result.

In such cases, coronary by-pass surgery is performed. During this procedure, a healthy vessel from another part of the body is transplanted into the heart.

This enables transport of nutrients and oxygen to the affected area to be resumed. The vessel is taken from the same individual to ensure tissue compatibility.

A bridge is constructed over the blocked coronary vessel and the flow of blood resumes through the by-pass vessels.

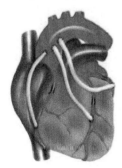

Coronary By-Pass Surgery

It describes the condition in which the coronary arteries supplying the heart become blocked. If insufficient blood is supplied to the heart, the individual feels a sharp pain in the chest.

The pain originates at the heart and spreads down the inside of the left arm. It is common for individuals with arteriosclerotic inflammation of the cardiac muscle to also have an accumulation of fat around the heart.

Smoking, as well as excessive consumption of tea, coffee and alcohol, increases the risk of infarction.

It may be initially treated by medication. Surgical treatment is given if the patient is considered to be at high risk.

2. Blood Vessels

The human circulatory system consists of arteries, veins and capillaries. The blood is sequentially pumped from the heart to the arteries, then into the capillaries and veins. It is returned to the heart via the veins.

a. The structure of vessels

Both arteries and veins are surrounded by a fibrous protective layer. This layer reinforces the strength of the arteries and veins against internal pressure created due to relaxation and contraction. The middle layer is composed of elastic fibers and smooth muscles, which contract and relax to facilitate the flow of blood.

In this layer there are also capillaries to supply nutrients to the veins and nerve fibres. The innermost layer of arteries and veins is composed of endothelium and provides a smooth and slippery surface to prevent friction.

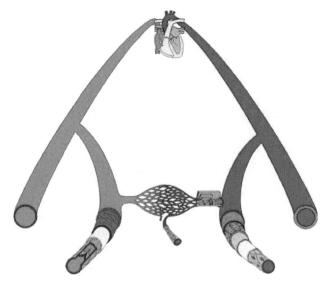

Figure: Shows the heart, arteries, veins and capillaries and their internal structure.

b. The types of vessels

There are three types of blood vessels.

Arteries   Capillaries   Veins

1. Arteries

These vessels transport blood from the heart to the different tissues of the body. All arteries, with the exception of the pulmonary arteries, carry **oxygenated blood**. The blood is pumped into the arteries by ventricular contraction. It is then forced into arterioles and capillaries. In comparison to veins, artery walls are stronger, thicker and more elastic. The elasticity of arteries produces tension in its walls.

The pulse is the rhythmic contraction and relaxation of arteries, coinciding with the contraction of the atria and ventricles during systole and diastole. The pulse rate is measured in the number of heart beats in a minute. The pulse rate is calculated by applying pressure on arteries in the body, such as at the wrist.

2. Capillaries

Non-muscular in structure, capillaries are located between arteries and veins and are composed of only endothelial cells, semipermeable, thin walled structures.

Figure: Blood vessels

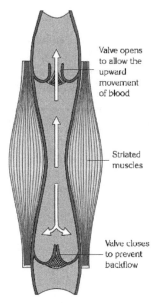

Valve opens to allow the upward movement of blood

Striated muscles

Valve closes to prevent backflow

Figure: The skeletal muscles surrounding the vessels contract, forcing the blood upward

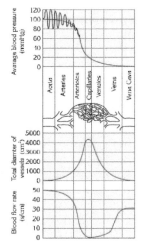

Figure: The above graph shows flow rate and diameter of arteries, capillaries and veins

Small molecules diffuse easily into or from the capillaries through this wall. Huge proteins such as plasma proteins cannot diffuse but may ooze into cells.

The total volume of blood in the capillaries is greater than in arteries and veins. Thus, blood flow at the capillaries decelerates due to friction. *Material exchange between blood and tissues is carried out at the capillaries.*

The density of the capillaries varies according to the type of tissue. It is high in skeletal muscles, very rare in adipose tissue and is not found at all in the lens of the eye.

3. Veins

Although they are structurally similar to arteries, they have a wider lumen (the part where the blood flows). Moreover, the walls of veins are thinner than those of arteries. However, the blood carrying capacity of the veins is greater than that of arteries. The veins force the blood to move in one direction due to the presence of valves, and are located close to the body surface. It can be observed that blood does not flow back if we squeeze the blood vessels in our hand. This provides evidence of the presence of valves in veins. By this method, blood at low pressure flows up to the heart. Any defect or enlargement of the valves causes an illness, known as **varix**.

c. The Blood Movement in Vessels

Blood pressure, skeletal muscles, body movements, valves in vessels and gravity are the factors which force the blood to move in vessels

Blood pressure is generated by the constriction of the ventricles. First, the blood passes through the arteries by ventricular contraction. Then it moves by means of the muscles in the vessels. Later, the blood is pumped upwards by the movement of skeletal muscles. The pulse rate in unit time is equal to the heart beat. The blood pressure at the entrance of the right atrium is less than 0 mm Hg. The blood therefore, is drawn into the atrium under a vacuum generated by the pumping of blood into the right ventricle.

The retrieval of blood to the heart is also influenced by other factors. Structurally, the veins are surrounded by skeletal muscles. Their contraction forces blood upwards, their valves preventing any back flow. The vein returns to its original diameter by relaxation of skeletal muscle.

Thus, blood pressure decreases in this portion of the vessel and is refilled with incoming blood from the capillaries. The activity of these muscles is important for the movement of blood against gravity. Standing for long periods can result in edema in the legs and feet. Adequate rest and movement help to prevent this.

Respiratory movements also affect the flow of blood in veins. During inhalation, air is taken into the lungs. In order to achieve this, the diaphragm enlarges and the chest expands. Thus the volume of the chest cavity increases and the pressure decreases both in this region and the veins. When the diaphragm flattens and chest pressure is reduced, vessels are stretched and this causes the movement of the blood.

d. Blood flow rate

The rate of blood flow is influenced by the diameter of the blood vessels and by blood pressure. It is most rapid in arteries, approximately 500 mm/sec and slightly slower in veins, approximately 150 mm/sec. However, it is slowest in the capillaries, 1 mm/sec. The movement of a fluid from a wide vessel to a narrow vessel increases the rate of flow. Conversely, the rate of blood flow decreases when it moves from a narrow vessel to a wider one. Therefore, blood flow is inversely proportional to the diameter of the vessel. Arteries, the widest vessels, separate into many narrow vessels known as arterioles. These are further divided into capillaries, which are narrower than the total volume of all the arteries. Thus, speed of blood flow decreases gradually from the heart to the capillaries. The total cross–sectional area of capillaries is 800 times as great as the total cross–sectional area of the aorta.

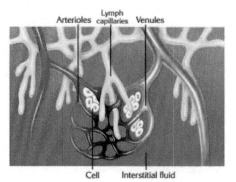

Figure: The location of arteries, veins, lymphatic capillaries, cells and interstitial fluid.

As a result, rate of blood flow in capillaries is 1/800 of that in the aorta. The capillaries converge to form small veins which then form larger veins, thus reducing the total volume. As a result, the rate of flow increases. The arterioles are directly connected to the veins in the skin, also known as **anastomoses**. The blood flows directly into the veins and, in this way, body temperature is regulated.

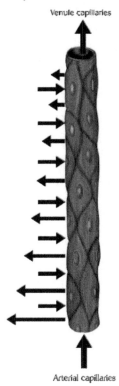

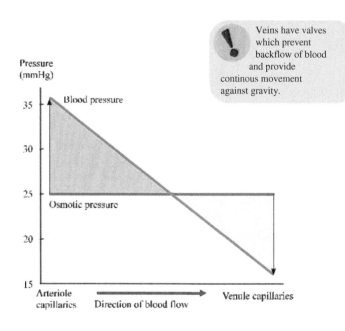

Veins have valves which prevent backflow of blood and provide continous movement against gravity.

Figure: The figures illustrate material exchange in the capillaries: Pressure is reduced from the arteriole end of the capillary to the veniole end. The osmotic pressure remains the same.

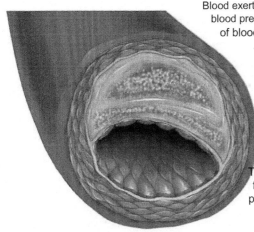

e. Blood Pressure

Blood exerts pressure on the walls of vessels during circulation. This is called blood pressure. It is constant and depends on two main factors: the amount of blood pumped, and the constriction of the ventricles.

The muscular structure of the vessel.

Blood pressure increases when the ventricles contract. It decreases during relaxation of the ventricles. In all mammals, the systolic pressure is approximately 120 mm Hg and the diastolic pressure is approximately 75–80 mm Hg. Blood pressure is at its maximum in the aorta, approximately 140 mm Hg, and at its minimum at the entrance of the right atrium, approximately 0 mm Hg.

The difference in the blood pressure of vessels is a prerequisite for the constant flow of blood. No movement of blood would be possible if the pressure were the same in all vessels. The blood pressure increases during physical exertion, but slows down during rest and sleep. Abnormal increase of blood pressure due to illness is termed hypertension, and abnormal decrease in blood pressure is called hypotension.

f. Material Exchange Between Cells and Blood

Cells and capillaries are embedded in interstitial fluid, which has an important function in material exchange between cells and blood. The thin walls of the capillaries allow the passage of small molecules from the blood to the interstitial fluid and cells. Conversely, some other molecules diffuse from the interstitial fluid into the capillaries. Metabolic waste from the cells diffuses into the venules via interstitial fluid. During material exchange, leucocytes and some proteins remain in the interstitial fluid. The lymphatic system then transports them into the circulatory system.

Only certain capillary beds (networks of many capillaries) are open at any given time. Shunting of blood is possible because each capillary bed has a channel that allows blood to go directly from arteriole to venule. Contracted sphincter muscles prevent the blood from entering the capillary vessels. After eating, for example, blood is shunted through the muscles of the body and diverted to the digestive system. This is why swimming after a heavy meal may cause cramping.

Types of Circulation

Pulmonary circulation

Pulmonary circulation of the blood occurs between the heart and the lungs. It is initiated with the contraction of the right ventricle and the pumping of deoxygenated blood into the pulmonary artery. Branches of the pulmonary artery transport blood into both lungs. In the lungs, CO_2 diffuses out of the blood into the lungs, while oxygen diffuses in. The oxygen-rich blood is then carried into the left atrium by the pulmonary veins.

Arteriosclerosis refers to the condition where the blood vessels become narrow and lose their elasticity. It is seen generally in men and wo-men over the age of 40. The vessels lose their elasticity due to a poor diet. Fats and Ca^{++} ions adhere to the walls of blood vessels,

causing narrowing. Subsequent disorders in the brain and heart then appear.

Deposition of fats and calcium may block the affected vessel and cause arterial bleeding, if the

condition remains untreated. Coagulation of the blood may cause paralysis. A low cholesterol / low salt diet is recommended for those suffering from this condition.

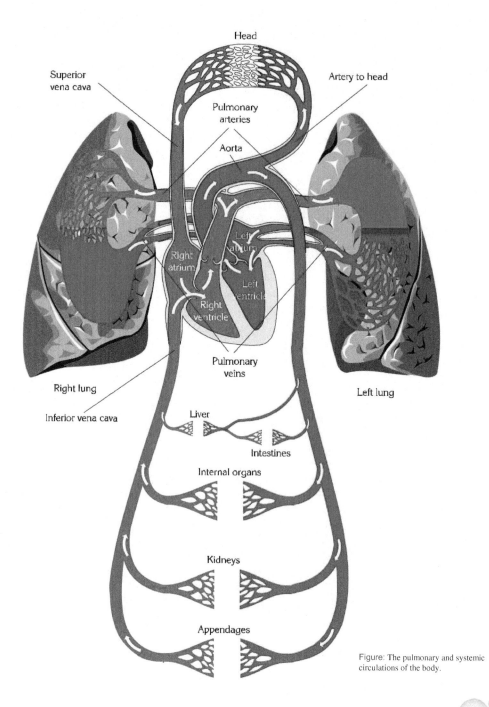

Head

Artery to head

Pulmonary
arteries

Aorta

Left
atrium

Right
atrium

Left
ventricle

Right
ventricle

Pulmonary
veins

Right lung

Left lung

Inferior vena cava

Liver

Intestines

Internal organs

Kidneys

Appendages

Figure: The pulmonary and systemic
circulations of the body.

Circulatory System

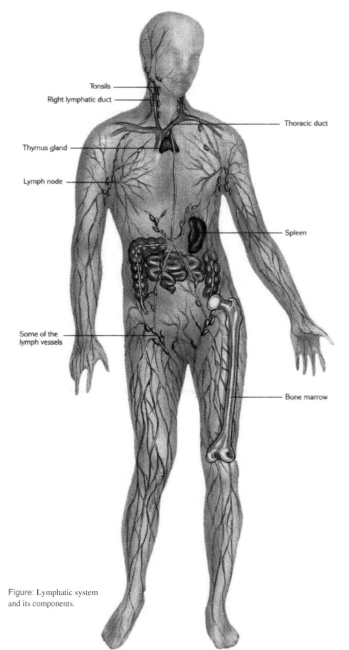

Tonsils

Right lymphatic duct

Thoracic duct

Thymus gland

Lymph node

Spleen

Some of the
lymph vessels

Bone marrow

Figure: Lymphatic system
and its components.

Systemic circulation

Systemic circulation occurs between the heart and all other parts of the body (except the lungs) where materials and gases are exchanged. It begins with the contraction of the left ventricle and the transport of oxygenated blood to the tissues via the aorta. The aorta then branches into the main vessels which carry blood into different parts of the body. The aorta descends and branches into the diaphragm and down into the coelom (body cavity). Its branches supply nutrients to the liver, intestines and other parts of the body. Nutrients and oxygen diffuse into the tissues while wastes and CO_2 diffuse into the blood. The deoxygenated blood is then transported by the superior and inferior vena cava into the right atrium.

Lymphatic Circulation

The blood circulatory system is aided by the lymphatic system in all vertebrates except fish. It is an independent system in human beings. The lymphatic system is composed of lymph capillaries, lymph vessels and lymph nodes. Lymph vessels are composed of transparent endothelial cells. Occasionally, they include bicuspid valves which direct the lymph to the heart. The lymph capillaries resemble open-ended tubes located between tissues. The fluid that has diffused from and has not been reabsorbed by the blood capillaries forms tissue fluid. Once fluid enters the lymphatic vessels, it is called lymph.

This system has three main functions: (1) lymphatic vessels take up excess tissue fluid and return it to the blood stream; (2) lymphatic capillaries, called lacteals, absorb fats at the intestinal villi and transport them to the blood stream; and, (3) the lymphatic system helps to defend the body against disease by means of lymph nodes, which filter microbes from the lymph and have certain roles in production of special immune cells.

The flow of lymph is similar to the blood flow in the veins. Movement of lymph is assisted by skeletal muscles, respiratory movements, valves, and pressure exerted by fluid flow. The rate of lymph flow within the vessels is slower than that of blood. Lymph nodes, ovoid or round small structures, are found at certain points along lymphatic vessels. The white blood cells in the lymph nodes are referred to as lymphocytes. Bacteria and other microbes are destroyed during the flow of lymph fluid in lymph node vessels. The lymph nodes swell in the presence of microbes within them.

 Microbes are filtered from lymph in lymph nodes.

Lymph Organs

These organs may be free or anchored to other organs. Lymph nodes, tonsils, thymus gland, Peyer's patches found within the intestinal wall, and the spleen are all organs of the lymphatic system.

3. Blood

Blood is a tissue and originates from the mesoderm of the embryo. It consists of 45% cells and 55% plasma.

There are approximately 15 liters of fluid in an adult human body. Blood comprises only 5 liters of the total volume of liquid. It can be easily separated by centrifugation due to a difference in density between plasma and its other components. Blood plasma has a density of 1.03 g/cm^3, while the other components of blood have a density of approximately 1.09 g/cm^3. Paraffin is smeared on the centrifugation tubes to prevent coagulation, and heparin or citrate is added to precipitate the calcium.

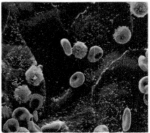

Figure: Blood is a special type of connective tissue.

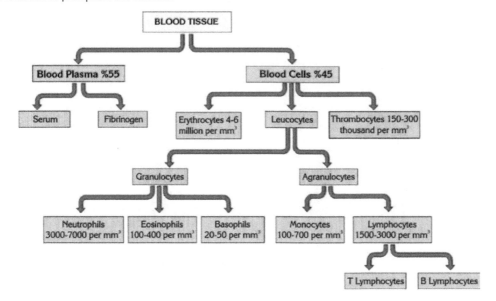

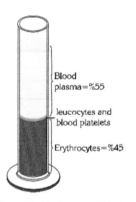

Figure: The blood includes 55% plasma, 44% erythrocytes and 1% other materials

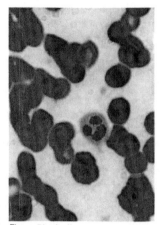

Figure: Blood cells.

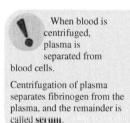
When blood is centrifuged, plasma is separated from blood cells.

Centrifugation of plasma separates fibrinogen from the plasma, and the remainder is called serum.

Functions of the Blood

1. **Nutrient Transport:** Nutrients, such as glucose, amino acids, vitamins, minerals and oxygen, are transported to cells by the blood. Metabolic wastes, such as carbon dioxide, urea, and water, are removed from cells and excreted.

2. **Hormone Transport:** Hormones secreted by the endocrine glands enter the blood and are transported to target cells or tissues.

3. **Homeostasis:** Blood helps maintain homeostasis by regulating pH at 7.4. It also regulates water and temperature levels.

4. **Immune response:** Invading viruses, bacteria and other foreign substances are phagocytosed by antibodies and leucocytes in the blood.

5. **Clotting:** During injury, blood loss is prevented due to the clotting capability of the blood.

The structure of blood tissue

As previously mentioned, blood consists mainly of cells and plasma and has a homogeneous appearance. If it is centrifuged without clotting, its homogeneous appearance is lost. Dense cells remain at the base of the tube and straw-colored plasma remains at the top. Serum may be obtained by the removal of fibrinogen from the plasma. The ratio of red blood cells to the total volume after centrifugation is termed the hematocrit. Erythrocytes comprise the lowest band (44%), leucocytes the middle band (0.05%), and thrombocytes form a very thin layer above the others (2.4%).

a. Plasma

Plasma constitutes 55% of the blood, of which 90-92 % is water, 7-9% plasma proteins, and the remaining 1% is amino acids, carbohydrates, lipids, hormones, urea, uric acid, lactic acid, enzymes, alcohol, antibodies, sodium, potassium, calcium, chloride, phosphate, magnesium, copper, iron, bicarbonate, iodine and other trace elements.

Some Important Proteins in Plasma: There are more than 70 different types of plasma proteins. Three of them are well-known and have important functions.

Fibrinogen is involved in blood clotting.

Albumin regulates osmotic pressure of the blood and interstitial fluid.

Globulins participate in the structure of antibodies.

Most blood proteins are produced by the liver The glucose level of blood is approximately 80-120 mg per 100 ml. If the amount of glucose decreases to 40 mg or below, hyperstimulation, fainting, shivering of the muscles and death, preceded by coma, occurs.

A diabetic is unable to control the level of glucose in the blood. If the glucose level decreases to a critical level, sugar must be ingested immediately. If the glucose level rises above a certain level, an injection of insulin is necessary to restore it to normal.

It is clear that glucose is used as a main energy source by the brain, so the amount of glucose must be kept constant by hormones in the blood. Besides all this, dissolved oxygen, nitrogen and carbon dioxide are also found in blood.

Blood cells are classified as erythrocytes (red blood cells), leucocytes (white blood cells) and thrombocytes (platelets).

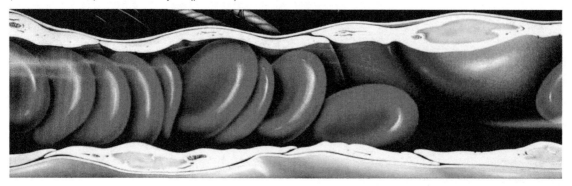

b. Blood Cells

1. Erythrocytes

Erythrocytes are 8 mm in length and 2 mm thick. There are approximately 5 to 5.5 million per mm^3 in the average male, and 4 to 4.5 million per mm^3 in the average female. Moreover, the number of erythrocytes varies considerably between humans living at different altitudes. At high altitudes, greater amounts of erythrocytes are expected due to a decrease in the partial pressure of oxygen in the atmosphere.

Mammalian erythrocytes are unique since they have no nucleus. However, the absence of a nucleus reduces their life span to between 80 and 120 days.

a. The Structure of Erythrocytes

Mature erythrocytes in mammals lack a nucleus, mitochondria, Golgi apparatus and endoplasmic reticulum. The lack of these organelles decreases the metabolism of the cell and increases their surface area. Half of the erythrocyte mass is available for oxygen loading.

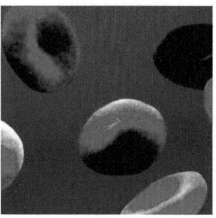

Figure: Red blood cells are disc-shaped, concave structures.

Hemoglobin consists of mainly two parts

A heme group

A globin group.

The heme group is an iron containing complex, whereas the globin group is composed of globular proteins (four polypeptide chains). Oxygen molecules bind weakly to the iron of the heme group. The globin group differs in each species of animal.

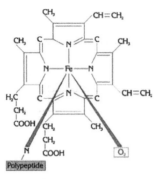

Figure: The heme group consists of oxygen and Fe atoms attached to globuline by N atoms to form a globular subunit. The four identical globular subunits associate to assemble a hemoglobin molecule.

Hemoglobin resembles chlorophyll in structure. However, iron is substituted for the magnesium found in chlorophyll. An erythrocyte contains approximately 265,000 hemoglobin molecules.

This structure increases the oxygen carrying capacity of blood. If a specialized physiological fluid were unavailable for oxygen transport, the amount of blood would need to be 75 times the normal volume, and the rate of blood transport would also have to be increased 75 fold.

Vertebrate hemoglobin is confined largely to the blood cells, while invertebrate hemoglobin is found in the plasma.

Figure: Shows the quaternary structure of the globular chain of hemoglobin. The oval structures are iron-containing heme groups. A hemoglobin molecule consists of two alpha (a) and two beta (b) chains.

b. Production of Erythrocytes

In the fetus, erythrocytes are produced by the liver and the spleen. They are also produced in the red bone marrow of the skeletal system, in the ribs and sternum. From the fifth month of development until the end of life, production occurs mostly in the marrow of the long bones. Erythrocytes proliferate from erythroblasts of red bone marrow. They lack hemoglobin, have a nucleus and divide rapidly, losing their nucleus, Golgi apparatus, mitochondria and other organelles after hemoglobin has been synthesized. They become characteristically concave and disk-shaped on both surfaces and are termed erythrocytes. This concavity facilitates their passage through the capillaries and increases their capacity to bind with oxygen due to the increase of their surface area. Another advantage of this concavity is the ability of the surface membrane to increase when needed. The erythrocyte membrane can swell under tension or in order to carry more O_2 and CO_2. Mature erythrocytes cannot respire aerobically since they lack mitochondria.

The most important vitamin in erythrocyte production is vitamin B_{12}. This vitamin stimulates cells to produce blood. Vitamin B_6 is responsible for recycling of erythrocytes, and vitamin B_{12} affects erythrocyte production in the bone marrow.

Human erythrocytes live for approximately 80-120 days in the circulatory system. This short life span can be explained by their lack of a nucleus and other organelles necessary for protein synthesis. Since erythrocytes cannot renew and regenerate themselves, old, worn-out erythrocytes are removed from the circulation by the spleen, liver and lymphoid nodules.

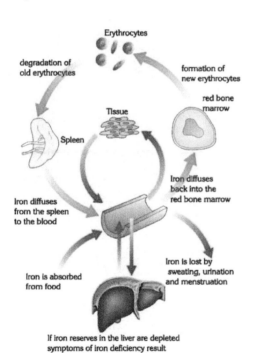

If iron reserves in the liver are depleted symptoms of iron deficiency result

Figure: Iron balance. Iron absorption from the intestine and iron excretion is almost equal in a healthy human being. The rate of absorption is the basic factor which maintains homeostasis. The degradation of erythrocytes provides 20 times the amount of absorbed iron. However, an iron deficiency for any reason may be satisfied by absorption.

Phagocytotic cells found in these organs digest erythrocytes and separate iron to be reused in new erythrocyte production in the red bone marrow. The porphyrine component of the heme group is converted to bilirubin and transported to the liver to form bile. This product is used by bacteria in the large intestine and is responsible for the color of the feces. When certain cells in the kidney do not receive enough oxygen, they produce a substance that combines with a plasma protein to form the hormone erythropoietin. This hormone stimulates red blood cell production in red bone marrow. Anemia is a reduction in red blood cells or in the amount of hemoglobin. These conditions decrease the amount of oxygen delivered to cells. and causes fatigue and lack of tolerance to cold. The most common cause of anemia is iron deficiency, treated by eating more iron-rich food.

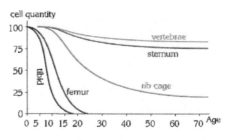

Figure: The graph above shows erythrocyte production in different bones.

2. Leucocytes (White Blood Cells)

Leucocytes are nucleated, spherical, white cells. They are also referred to as white blood cells due to their color. The number of leucocytes in the blood of a healthy person is approximately 6000/mm^3. Leucocytes may be found in both blood and interstitial fluid. They can attach themselves to the internal surface of the endothelium or vessel wall and move against the flow of blood. They can also cross the capillary walls in interstitial fluid. There are three main types of leucocytes and all of them are produced both in red bone marrow and in lymph nodes.

a. Types of Leucocytes (WBC)

Leucocytes (WBC) are divided into two main groups according to their size, the shape of their nucleus and granules found in their cytoplasm.

1. Granulocytes

Granulocytes are formed in red bone marrow. They are phagocytotic cells. They have segmented nuclei and granules within their cytoplasm. These granules are known as peroxidase enzyme-carrying lysosomes. The number of lobes of the nucleus is related to the age of the cell. When the granules mature, they lose their ability to divide.

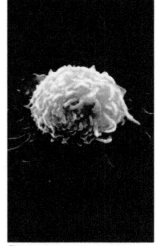

Figure: A typical white blood cell.

Neutrophils: These cells constitute 60-70 percent of leucocytes. They are 9-12 μm in diameter. Their nucleus also consists of 2-5 lobes linked to each other by chromatin threads. Neutrophils digest non-functional, worn out cells and bacteria by phagocytosis. The chemical compounds of inflamed tissue attract neutrophils by chemotaxis.

Eosinophils: Eosinophils constitute only 1-4 percent of leucocytes. They are 10-12 μm in diameter and their nucleus has only two lobes. Granules containing the enzyme peroxidase are found in the cytoplasm. Their phagocytotic activity, however, is weaker than the other leucocytes. Eosinophils are sensitive to allergens. The number of eosinophils increases when an allergen is present.

 Certain cells in kidneys are highly sensitive to O_2 insufficiency.

When O_2 is not sufficient these cells trigger a hormone secretion, erythropoitein, and red bone marrow is stimulated for more red blood cell production.

Figure: Typical white blood cells.

Basophils: Basophils form 0.5% of leucocytes. They are 8-11 μm in diameter. The nucleus of a basophil can be lightly stained by a basic dye. It is larger and has fewer lobes. A basophil cannot phagocytose particles, but can secrete heparin and histamines. Where there is long-term inflammation, their numbers increase.

2. Agranulocytes

The nucleus of an agranulocyte lacks lobes and is partly spherical in shape. It also differs from a granulocyte by its ability to divide. Most agranulocytes, lymphocytes and monocytes are produced in the lymph nodes, the spleen and thymus.

Monocytes constitute 2-8% of leucocytes. They are 12-20 μm in diameter and are therefore the largest type of leucocyte. They have a distinctive kidney-shaped nucleus and a greater amount of cytoplasm. The ability to phagocytose is highly developed. They differentiate into macrophages which can move between the blood capillaries and connective tissue and can engulf approximately 100 bacteria at once. They play an important role in the control of long-term bacterial infections and in the cleansing of injured tissue.

Lymphocytes: The second most abundant leucocytes are lymphocytes in the form of neutrophils. Lymphocytes constitute 20-35 percent of all leucocytes. Lymphocytes are the smallest type of leucocyte, each cell is 8 μm in diameter. Its large nucleus almost fills the cell and the cytoplasm is much reduced.

The ability of lymphocytes to differentiate into other cell types is of particular interest. They first swell to form monocytes, and afterwards differentiate into macrophages. Lymphocytes can be found in all tissue types except nervous tissue. They are abundant in the lymph system and in the blood. Lymphocytes have no ability to phagocytose and show no chemotaxis. The main site of production is the thymus gland. In the fetus, lymphocytes are produced in the thymus gland, then enter other lymph organs, namely the spleen, lymph nodes and red bone morrow. There they continue to multiply.

According to their site of production, they are divided into two groups: T and B lymphocytes. T lymphocytes are produced in the thymus gland, and B lymphocytes in birds are produced in the bursa fabriceus, a lymphoid tissue. In mammals, B lymphocytes are produced in the Payer's patches of the intestine, the tonsils and the mucosal layer of the caecum. The organ of production confers either T or B codes on its lymphocytes.

In the event of infection by microorganisms, lymphocytes are converted into aggressive defense cells, equipped to fight against attack. Lymphocytes have different defense strategies. They can attack whole foreign bodies and also fight at a molecular level by producing antibodies. Defence at the antibody level against harmful microorganisms is known as humoral immunity.

Defence against complete microorganisms is known as cellular immunity. T-lymphocytes are effective in cellular immunity, whereas B lymphocytes are active in humoral immunity. Once in the plasma at the site of infection, these cells secrete antibodies against foreign antigens.

READ ME

It is considered as a disorder rather than a disease, and results from a decrease in erythrocyte number or a decrease in the amount of hemoglobin in each erythrocyte, or a decrease in both. Excess blood loss, malaria, sickle-cell anemia, accumulation of heavy metals causing disorders in the liver and spleen, iron and B-12 deficiency all cause anemia. It is also seen in young children fed only on milk that is iron deficient. Characteristic symptoms of anemia include accelerated heart beat, degradation of tissues and reduced transport of oxygen in the blood.

In such cases, a course of Vitamin B-12 should be advised to a patient.

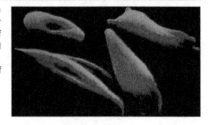

Sickle-cell anemia

This condition is seen in 0.3% of black Americans and Africans. An abnormality in the globin region of erythrocytes causes bending of cells into a sickle shape. When these hemoglobin molecules release their oxygen, the hemoglobin is crystallised, changing the shape of the cell to an "S" form. Such crystallization causes difficulties for blood to flow into the tissues. Thus, the oxygen supply to the tissues is reduced. Consequently, a reduction in oxygen causes further sickle-cell formation. As a result, the rate of malformation increases rapidly, and may cause death in a few hours.

Polycythaemia

Symptoms of this condition appear when the number of erythrocytes increases to 11-15 million/mm^3. The risk of this disorder developing is particularly high when there is either rapid heat loss from the body or severe diarrhea. The number of erythrocytes remains constant, although the volume of blood is reduced and blood vessels may become blocked due to increased viscosity. Heart function subsequently becomes difficult in such cases.

READ ME *BITES*

The most potentially dangerous are bites from snakes, and stings from scorpions. After being bitten/stung, it is important to record a description of the type of snake or scorpion so that the correct anti-venom can be later administered. The symptoms and signs of a bite are one or two small, puncture-like wounds in the skin and swelling around the affected area. The casulty may show signs of shock as well as disturbed vision, nausea and difficulty in breathing. It is important to restrict the spread of poison through the circulation. To achieve this, an incision can be made in the region of the wound using a sterile knife and the poision sucked out. A tie or belt should then be tied tightly above the affected limb to isolate the poison, The casualty should lie down and be advised not to move until he can be taken to hospital. The tie or belt above the wound should be loosened every half hour to allow blood circulation to the affected area.

Animal Bites

A bite from an animal carries both the risk of infection from the germs on the teeth being injected deep into the tissues and, more seriously, from rabies. The saliva of infected animals contains the rabies virus, which will cause death if prompt action is not taken.

Animal bites usually cause damage to the tissues with inevitable blood loss in the case of multiple bites. The affected area should be thoroughly washed with soap and water and then covered with a sterile dressing. To prevent blood loss from deep bites, pressure should be applied to the affected region and it should be kept in an elevated position. The animal should be isolated, if possible, so that it can be examined for rabies. In any case, the first of a course of injections against rabies should be administered promptly.

Septicemia

This serious disease can affect those whose immune system has been weakened due to, for example, a previous infection. Fast replicating meningococcic bacteria invade the body through a wound or, more commonly, by the respiratory tract. They penetrate the blood vessels, spreading throughout the circulation to attack tissues of the body of high inflammatory risk such as the tonsils, gall bladder, kidneys and uterus.

The first symptoms of infection are fatigue and numbness in the affected areas.

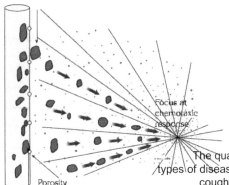

Focus at chemotaxic response

Porosity

b. The Chemotaxic Response of Leucocytes

Most chemical compounds in tissues affect leucocyte activity. Neutrophils and monocytes respond to chemicals, particularly polysaccharides, produced by inflamed or infected tissue, by moving to the site of infection. Many bacterial toxins also stimulate a chemotaxic response in leucocytes. Their stimulation is dependent on the density of the active chemical source. Leucocytes move toward areas of high source chemical density by forming pseudopodia. However, in some cases leucocytes move in the opposite direction, to escape.

The quantity of white blood cells in the body can be used to identify certain types of disease. For instance, the quantity of lymphocytes increases in whooping cough, severe anemia, skin deformation due to over exposure to the sun, tuberculosis and chronic illness. The number of monocytes increases when an individual is suffering from typhoid or malaria. A high number of neutrophils indicates bacterial infections such as pneumonia. The amount of eosinophils increases during allergic diseases, fever, asthma and some skin disorders.

 The quantity of white blood cells in the body can be used to identify certain types of disease. For instance, the quantity of lymphocytes increases in whooping cough, severe anemia, skin deformation due to over exposure to the sun, tuberculosis and chronic illness. The number of monocytes increases when an individual is suffering from typhoid or malaria. A high number of neutrophils indicates bacterial infections such as pneumonia. The amount of eosinophils increases during allergic diseases, fever, asthma and some skin disorders.

3. Platelets

 Blood platelets, or thrombocytes, number approximately 250,000/mm^3 of blood. They are 2-4 µm in diameter, non-nucleated and white in color. Platelets have characteristic ray-like surface projections that play an important role in blood clotting. The blood platelets of hemophiliacs lack these surface projections.

Thrombocytes originate from large, nucleated cells in the bone marrow and are formed from the degradation of phagocytes in the lungs. Their life span is only approximately 8 days.

Blood clotting

They play a vital role in preventing blood loss from the body and maintain hemostasis by blood clotting, thus helping to prevent the loss of large volumes of blood.

True blood clotting can be observed in crabs and all vertebrates. If a blood vessel is damaged or punctured, thrombocytes produce serotonin, which shrinks the area of injury, thus preventing the flow of blood in the affected capillaries. This also retards the flow in the veins. A plug of platelets is formed, followed by a blood clot. Clotting is made possible by globular proteins, synthesized in the liver, known as fibrinogen and thrombogen.

Thrombocytes secrete serotonin before coagulation occurs. They also secrete the enzyme thrombokinase upon contact with the air. This enzyme, thrombokinase, converts thrombogen into thrombin in the presence of Ca^{++} ions. The reaction between thrombin and fibrinogen produces fibrin threads, which stick to the damaged surface of the blood vessel and form the webbing of the clot. These threads also trap blood cells and platelets, which help to strengthen the clot.

After the damaged area of the vessels has been closed, the plasma produces an enzyme. This enzyme inhibits clotting in blood vessels unaffected by injury. If the body takes no steps to localize the clotting response, the vessels may be blocked and death can occur. An effective vitamin in blood clotting is vitamin K. The function of vitamin K is to convert prothrombin into thrombogen in the liver. Heparin inhibits clotting in the vessels.

Blood typing

Blood typing involves two types of molecules called antigens and antibodies. Organisms have some proteins called antigens. When they enter the bodies of other organisms, they are not recognized, and special molecules, called antibodies, are produced to destroy the foreign antigens. The membrane of red blood cells contains antigens. The plasma of a recipient of transfused blood may contain antibodies that will react against them. Such reactions are life threatening and, much more than the donor, the recipient must be careful not to receive blood containing foreign antigens.

ABO System

There are many different systems for typing blood, but the most common one is the ABO system. On the membrane of a red blood cell, only antigen A, only antigen B, both antigen A and B, or no antigen may be found. In the simplified ABO system there are four possible blood types:

A blood group: Red blood cells carry only antigen A. The plasma of a such person has antibody b.

B blood group: Red blood cells carry only antigen B. The plasma of a such person has antibody a.

AB blood group: Red blood cells carry both antigen A and antigen B. Since both of the antigens are present, there is no antibody in the plasma.

O blood group: Red blood cells carry none of the antigens. Since both of the antigens are foreign to the organism, plasma has both antibody a and antibody b.

If a foreign blood antigen enters the body of an organism, the antibodies will be activated and agglutination, or clumping, of red blood cells will occur. Agglutination of red blood cells can cause blood to stop circulating in small blood vessels, and this leads to organ damage and death. For a recipient to receive blood from a donor, the recipient's plasma must not have an antibody that causes the donor's cells to agglutinate.

O type blood can be given to all others since it has no antigens to be destroyed (universal donor).

AB type blood has no antibodies in the plasma, thus can receive any type of blood (universal recipient).

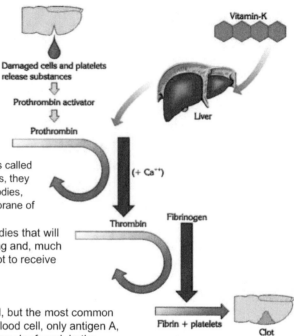

Damaged cells and platelets release substances

⇩

Prothrombin activator

⇩

Prothrombin

(+ Ca^{++})

Thrombin   Fibrinogen

Fibrin + platelets   Clot

Vitamin-K

Liver

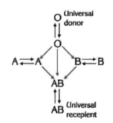

O Universal donor

A ⇄ A   B ⇄ B

AB

AB Universal recipient

Figure: Blood transfusion.

Rh system

Another important antigen in blood typing is antigen D. Individuals with this antigen are Rh (+), while the ones not possesing this antigen are Rh (-). Rh (-) individuals normally do not have antibody d against Rh factor, but they may make it when exposed to the Rh factor (antigen D)

During pregnancy, if the mother is Rh (-) and the father is Rh (+), the child may be Rh (+). The Rh (+) red blood cells may begin leaking across the placenta into the mother's circulatory system, as placental tissues normally break down before and at birth. Antigen D taken from babies will cause formation of antibody d.

In this or a subsequent pregnancy with another Rh (+) baby, antibody d may cross the placenta and destroy the child's red blood cells. Due to red blood cell destruction followed by heme breakdown, bilirubin rises in the blood. Excess bilirubin can lead to brain damage and mental retardation, or even death.

This problem can be prevented by giving Rh (-) women an Rh immunoglobin (antibody d) injection no later than 72 hours after giving birth to any Rh (+) baby. These antibodies will destroy any of the child's red blood cells in the mother's blood, before the mother's immune sytem recognizes them.

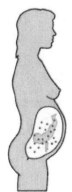 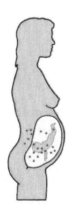  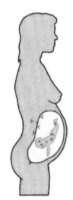

| Baby is Rh (+) Mother is Rh (-) | Baby's antigens D enters mother's body through placenta | Mother's immune system produces antibody D | Antibodies attack red blood cells of baby |

Young couples must know their blood groups and learn if they have any risks before they marry and have babies.

Treatment of some health problems is very simple, but only if we are aware of them.

IMMUNOLOGY

Organisms resist pathogenic elements and foreign substances in their environment. Despite all these barriers, a few pathogenic elements may actually penetrate the body's circulatory system. In response, the body produces some substances that react to these alien cells.

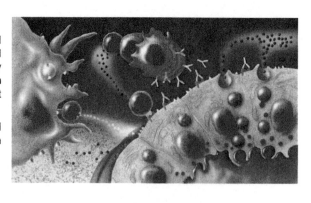

Immunology is the study of the strategies and mechanisms that the body uses to rid itself of foreign substances.

1. Organs Of The Immune System

a, The Spleen

The spleen is an organ of the immune system and is located directly beneath the diaphragm in the upper-left section of the coelom. It is approximately 200 g in weight.

The spleen is involved in:

degradation of old and dead erythrocytes.

storage of blood as a reserve in the event of any shortage.

production of lymphocytes active in the defense of the body. Both the spleen and liver work cooperatively in production.

production of fetal blood until birth. It is then produced by red bone marrow.

The functions of the spleen can be performed by other organs if the spleen has to be removed from the body due to injury. Therefore its role is not vital.

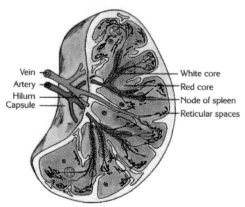

Figure: The spleen is an organ of the immune system.

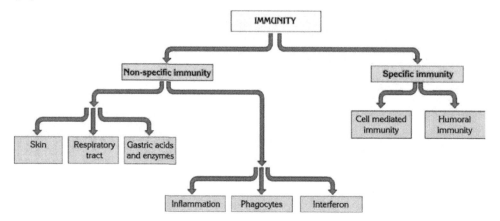

Circulatory System

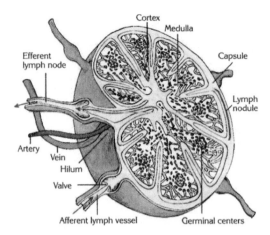

Figure: In the figure, the direction of lymph flow is indicated by arrows. The lymphocytes, monocytes, macrophages and antigens may make contact with each other by this movement. Thus, their defense function is carried out

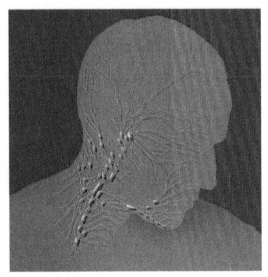

Figure: Lymph nodes found in the upper portion of the body.

b. Lymph nodes

Lymph nodes are surrounded by a capsule composed of connective tissue. Lymph nodes are engaged in both blood production and body defense by capturing microbes. In the event of major infection, the lymph nodes can become enlarged and can be felt during examination. The locations of the lymph nodes in the body are the armpits, groin, neck, elbow and chest. Lymph fluid is transported into the lymph nodes by divergent vessels. Lymphocytes are filtered at these nodes and the fluid enters the circulatory system via the blood vessels.

c. Glottis

The glottis forms a capsule around the pharynx. Its size in infants is notably larger than in adults. The direction of the lymph flow at the glottis is towards the nodes of the neck. Thus, in the event of infection, these structures are frequently inflamed.

d. Mucosal nodes:

The mucosal tracts of the body contain lymph islets. They are particularly abundant in the intestinal mucosa. Thus, lymphocytes are produced in these structures.

e. Reticulo-endothelial System

The reticulo-endothelial system consists of a rich network of blood capillaries and cells of fibrous tissue. *The reticulo-endothelial system is made up of Kuppfer cells of the liver and reticular cells of red bone marrow.* This system defends the body against hazardous substances by initiating antibody release and phagocytosis. The reticular cells, or histocytes, are stimulated by toxic or mechanical means and reach their target area by ameboid motion. They then phagocytose the invading microorganisms. The stem cells of red bone marrow are involved in the production of lymphocytes.

f. Thymus

The thymus is located under the thyroid gland in the chest cavity. It is surrounded by a capsule composed of connective tissue. The thymus gland defends the body against infection by producing lymphocytes just before and after birth. Reticular cells, macrophages and lymphocytes are located in its lobes.

As mentioned previously, immunity is the recognition and removal of molecules foreign to the body. It is categorized as non-specific (inherited) immunity and specific immunity.

2. Acquisition of Immunity

Active immunity: In active immunity, the individual alone produces antibodies against an antigen. Active immunity sometimes develops naturally after a person is infected with a pathogen, or an individual can be artificially immunized against the pathogens. Immunization involves the use of vaccine, substances containing an antigen to which the immune system responds. Active immunity is long-lived.

Passive immunity: In passive immunity, the individual is given prepared antibodies. These antibodies are not produced by the individuals B lymphocytes. Passive immunity is short-lived. Ready antibodies can be taken from a person who recovered from the illness, or passed from mother to baby through placenta (before birth), or in the milk during breast-feeding.

3. Types of Immunity

Immunity is maintained by two pathways:

non-specific immunity

specific immunity

a. Non-specific Immunity

Barriers nonspecifically prevent microbes from entering the body. That is, they block all microbes, without distinguishing whether they are harmful or not. It is maintained by interferon, phagocytosis, skin, tears and sweat, gastric juices, hair, inflamation and mucus in the respiratory tract

The skin, acting as a mechanical barrier, protects human beings from external bacteria and harmful microorganisms. Tears and sweat, since they have a high salt concentration, eliminate some bacteria. Microbes ingested in food are destroyed in the stomach by gastric juices, which include HCl and enzymes. Dust particles and other airborne substances taken in during inhalation are filtered in the nasal cavity. Germs that succeed in entering the trachea are trapped by a mucous layer, while pathogens that manage to reach the alveoli are eliminated by macrophages. Inflamation is a nonspecific defense that makes an environment hostile to microorganisms. If the skin is inflamated, first phagocytes infiltrate the injured skin, attacking entering bacteria. Plasma accumulates at the wound site, diluting toxins that bacteria secrete and bringing antibacterial substances to the site. Increased blood flow warms the area, turning it swollen and red.

In nonspecific immunity, the body also uses certain chemicals called interferons. Interferon is the term given to protein molecules which are produced by host organisms (cells that are infected by microbes) in response to infection by a pathogenic virus. Their function is to deactivate viruses. They are nonspecific to viruses. However, they do occur in different forms.

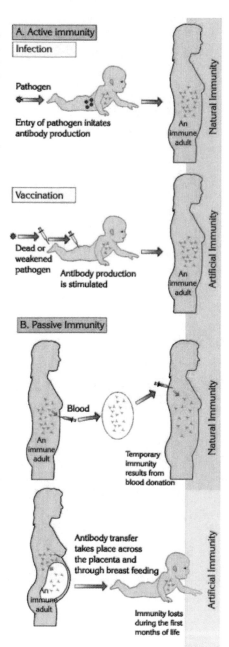

A. Active immunity
Infection

Pathogen

Entry of pathogen initates antibody production

Natural Immunity

An immune adult

Vaccination

Dead or weakened pathogen

Antibody production is stimulated

An immune adult

Artificial Immunity

B. Passive Immunity

An immune adult

Blood

Temporary immunity results from blood donation

Natural Immunity

Antibody transfer takes place across the placenta and through breast feeding

An immune adult

Immunity losts during the first months of life

Artificial Immunity

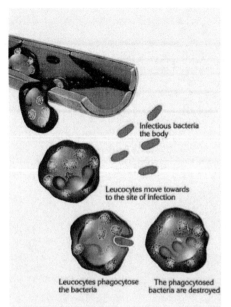

They act to prevent viral reproduction. In order to achieve this, they attach themselves to infected host cells and stimulate the production of enzymes by the host cell to block protein synthesis of the virus. These extremely rapid and accurate events continue until the immune system is activated. In addition, interferon stimulates lymphocytes to degrade infected or cancerous cells.

Recombinant DNA Technique

The gene responsible for the production of interferon is placed into the genome of a bacterium. Since the rate of bacteria fission is rapid, large amounts of interferon may be synthesized in a short period of time.

Recent investigation has shown this technique has a useful application in cancer research. In recent days, the recombinant DNA technique has been used in interferon production in the pharmaceutical industry.

A fever also protects nonspecifically. It begins when viruses or bacteria affect some white blood cells. Some of these cells produce interleukin proteins, which affect the thermoregulatory center in the hypothalamus to maintain a higher body temperature.

The heat kills some invading microbes directly. It also prevents microbial growth indirectly, because higher temperature reduces the iron level in the blood.

Because bacteria and fungi need more iron as the temperature rises, a fever may stop their growth. Phagocytotic cells also attack microbes better when the temperature rises.

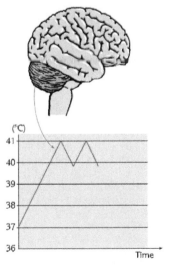

Figure: The graph indicates the alteration of temperature during inflammation. Increase of body temperature helps the body to destroy microbes.

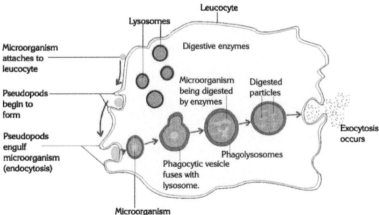

Figure: Stages of phagocytosis

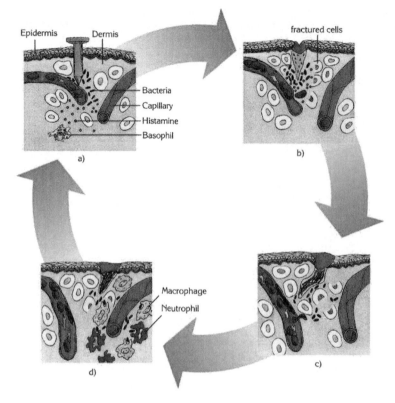

Figure:
a. Any break in the skin due to injury can cause the invasion of bacteria, resulting in cell fracture and the accumulation of histamine.
b. The blood vessels in the affected region swell, increasing the rate of blood flow. This results in the accumulation of interstitial fluid in this region.
c. A clot is formed and the normal and fractured cells separate. The neutrophils attach themselves to the walls of the blood vessels.
d. As new tissue is generated, the clot is degraded.

b. Specific immunity

It is based on production of a different type of antibody against each different type of microorganism antigen. T and B lymphocytes have the main responsibility in specific immunity. However, the first cell type that responds to infection is the macrophage, which then activates lympocytes. It is maintained by two pathways: humoral immunity and cell mediated immunity.

1. Humoral immunity:

B lymphocytes secrete antibodies into the blood stream ("humor" means fluid). The response begins when T lymphocytes react to a foreign antigen. The stimulated B cell then divides, producing many identical cells that can identify the foreign antigen (specific protein of the foreign organism).

These B cells then change into plasma cells or memory cells. A memory cell secretes up to 2000 identical antibodies (structures used as weapons against foreign antigens) per second. These antibodies surround, bind, and inactivate foreign antigens. Memory B lymphocytes respond to the antigen quickly and forcefully. If the same type of antigen later enters the body again, these memory

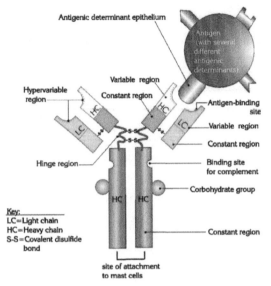

Key:
LC=Light chain
HC=Heavy chain
S-S=Covalent disulfide bond

Labels in figure: Antigenic determinant epithelium; Antigen (with several different antigenic determinants); Variable region; Hypervariable region; Constant region; Antigen-binding site; Variable region; Constant region; Hinge region; Binding site for complement; Carbohydrate group; Constant region; site of attachment to mast cells

Figure: An antibody consists of two long and two short chains. The upper variable portions are specialized for the antigens which attach to them.

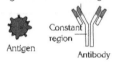

Antigen surface   Variable region
Constant region
Antigen   Antibody

Antigen-Antibody complex

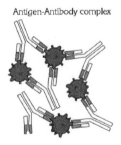

Figure: The structure of an antigen, antibody, and antigen-antibody complex

cells will be used, and a much quicker response will be seen. A first reaction to a particular foreign antigen is a **primary immune response**. Memory cells formed from some of the activated B cells provide a quick **secondary immune response** when the same antigen enters the body again.

1. 1. Antigens:

Antigens consist of foreign substances that initiate the formation of antibodies when they enter the human's or other animal's body. Antigens facilitate the formation of antibodies and also react with them, both inside and outside the body.

1. 2 Antibodies

All vertebrates can synthesize antibodies. They are formed by stimulation of the antigen and react with them. They are also known as immunoglobulins. Antibodies are synthesized by plasma cells, which are formed by B lymphocytes. All antibodies are characterized by antigen recognition sites on their surfaces.

The structure of antibodies: Antibodies structurally are globular proteins known as immunoglobulins. Each immunoglobulin is composed of four chains of amino acids, bonded together by a disulphate bond. The sequence of amino acids constituting an immuno-globulin is unique.

Antigen-Antibody reactions: Antibodies are structurally peculiar to their antigens. A compatible antibody and antigen form an antibody-antigen complex, which functions as a lock and key, each antibody binding specifically with its antigen type. Viruses and bacteria, upon penetrating the body's defenses, disrupt the metabolism of the organism by reproducing and releasing toxic substances. The disease-causing organism is referred to as the pathogen, and its ability to cause disease is called virulence. The reproduction of viruses and bacteria is termed infection. The carrier organism is initially unaffected by pathogens while they are reproducing inside it. The following events are the results of antigen-antibody reactions. Generally, antibodies make direct contact with antigens. Four different results of these reactions are as follows;

Agglutination: antibodies and antigens react and are inactivated.

Precipitation: the antibody-antigen complex precipitates out of the solution.

Neutralization: the antibodies block the toxic portion of the antigen, thereby preventing its harmful effects.

Lysis: Antibodies cause the rupture of the plasma membrane of the antigen, so inactivating it.

The complementary system is clusters of enzymes located in plasma in an inactive form.

1. 3. Immunoglobulins

There are five different types of immunoglobulins: IgG, IgM, IgA, IgD and IgE.

IgG is abundant in human serum and comprises approximately 80% of it. It is the only immunoglobulin which passes across the placenta from the mother to the fetus. It plays an important role in protection against viruses. Immunoglobulin A, D and M are antibody types which are abundant in plasma. IgE is of least quantity and plays an important role in allergies of the skin and other tissues.

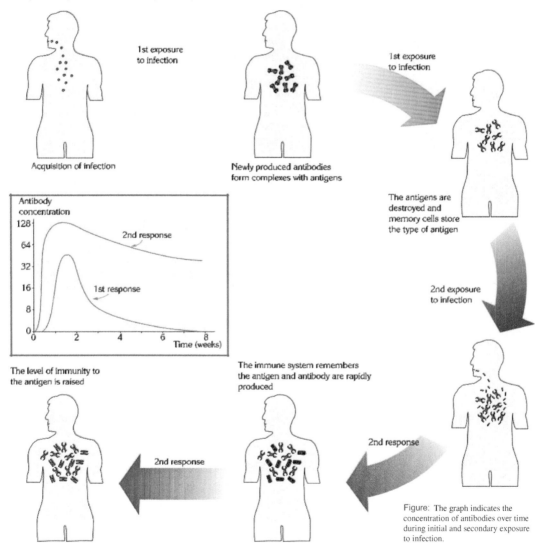

1st exposure to infection

Acquisition of infection

Newly produced antibodies form complexes with antigens

1st exposure to infection

The antigens are destroyed and memory cells store the type of antigen

2nd exposure to infection

The level of immunity to the antigen is raised

The immune system remembers the antigen and antibody are rapidly produced

2nd response

2nd response

2nd response

Figure: The graph indicates the concentration of antibodies over time during initial and secondary exposure to infection.

Circulatory System

CLASSIFICATION OF THE IMMUNOGLOBULIN	
Immunoglobulin	*Location*
IgG	All body fluids, especially serum
IgM	Serum
IgA	Glands and serum
IgE	Mast cells
IgD	At the surface of the lymphocytes

2. Cell mediated Immunity (cellular immune response)

T lymphocytes are involved in cell mediated immunity. It is called cellular because the T cells move to where they are needed (B cells don't change their position, they only produce antibodies and give them to the blood stream). In the thymus, T cells mature and gain the ability to recognize different types of surface antigens. There are several types of T lymphocytes and each one has a distinct function.

Cytotoxic T cells kill cells on contact. Helper T cells produce chemicals (lymphokines) and stimulate other immune cells. Suppressor T cells suppress the immune response. Memory T cells remain in the body to provide long-lasting immunity.

Figure: Illustrates the formation of T & B lymphocytes in bone marrow. The lymphoblasts reach the thymus gland by blood circulation. They are carried into the lymph nodes as T-lymphocytes. Inactive T-lymphocytes are activated when they react with antigens, thus they maintain cell-mediated immunity. B-lymphocytes are involved in the maintenance of humoral immunity and they originate from the bone marrow in mammals.

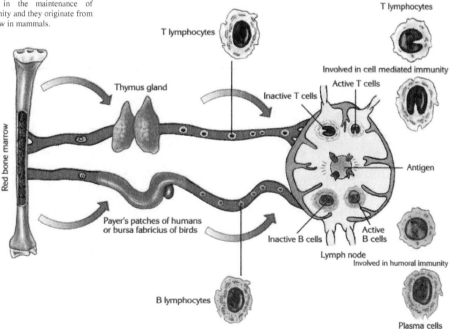

HUMAN BIOLOGY

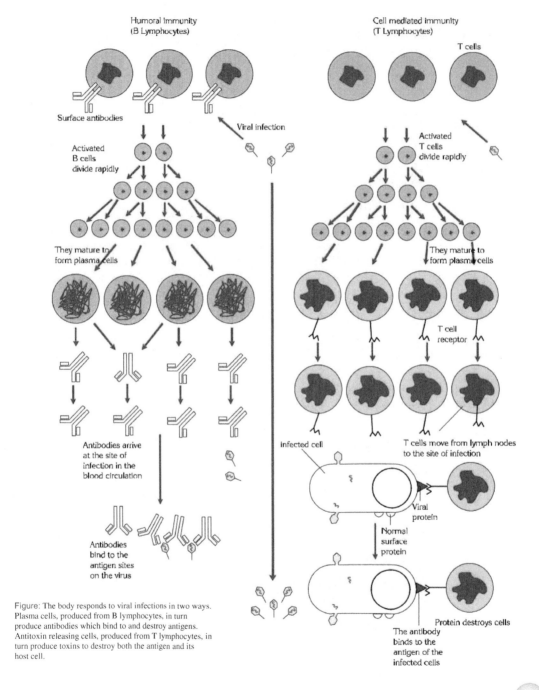

Humoral immunity
(B Lymphocytes)

Cell mediated immunity
(T Lymphocytes)

T cells

Surface antibodies

Viral infection

Activated
B cells
divide rapidly

Activated
T cells
divide rapidly

They mature to
form plasma cells

They mature to
form plasma cells

T cell
receptor

Antibodies arrive
at the site of
infection in the
blood circulation

infected cell

T cells move from lymph nodes
to the site of infection

Viral
protein

Normal
surface
protein

Antibodies
bind to the
antigen sites
on the virus

Protein destroys cells

The antibody
binds to the
antigen of the
infected cells

Figure: The body responds to viral infections in two ways.
Plasma cells, produced from B lymphocytes, in turn
produce antibodies which bind to and destroy antigens.
Antitoxin releasing cells, produced from T lymphocytes, in
turn produce toxins to destroy both the antigen and its
host cell.

Circulatory System

143

Inhalation of pollen grains

⬇

Allergens on the pollen grains stimulate secretion of IgE from plasma cells

⬇

IgE binds to the cells of the respiratory tract

⬇

IgE binds to the allergens

⬇

Mast cells secrete histamine and other chemicals

⬇

The respiratory tract swells   The capillaries dilate

⬇

Allergic responses observed are odema, constriction and red spots in the respiratory tract

Figure: The stages of development of an allergy.

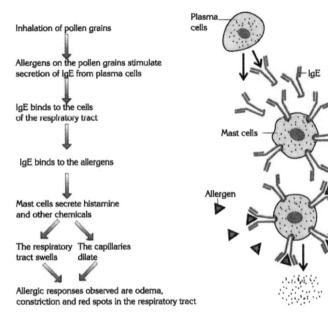

The most effective way to fight virulent disease is to confer recognition of the antigen to the body. Vaccination is a way of equipping the body to fight against disease.

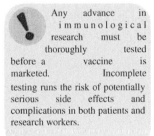
Any advance in immunological research must be thoroughly tested before a vaccine is marketed. Incomplete testing runs the risk of potentially serious side effects and complications in both patients and research workers.

4. Allergy

All allergies can be described as a type of response by the immune system to infection from disease. The symptoms of an allergy originate from the activity of antigens and antibodies in the lymphatic system.

Antigens react with these antibodies and trigger a sequence of events that produces allergy symptoms. Histamine and histamine-like substances are produced, causing the enlargement of capillaries and the contraction of smooth muscle.

The enlargement of capillaries results in a flow of plasma from the blood to the intercellular fluid. As a result, allergic symptoms, such as edema and high fever, appear.

A few bacteria, such as tuberculosis bacillus, produce an allergic response. These bacteria are called allergens.

5. Immunologic Tolerance

An organism must be able to recognize its own proteins as non-threatening and be able to show immunological tolerance to them. The limits of immunological tolerance are determined towards the end of embryonic development, and stabilize after birth. During embryonic development, no response is made to foreign proteins, and they are accepted as one's own. If some foreign proteins are injected during adolescence, individuals tolerate such particles. The antibodies in the blood are created during childhood. For example,

Anti-B against Antigen-B,

Anti-A against Antigen-A.

Antigen D against the Rhesus factor appears during an Rh (+) injection in Rh (-) individuals.

During organ transplantation, tissue compatibility should be determined. Otherwise the organism will recognize the organ as hostile and produce antibodies to destroy it, rejecting the transplanted organ. Such a severe immune response is depressed by drugs to ensure compatibility.

6. Vaccines

Before modern medicine, the effect of the immune system to diseases such as small-pox and measles was recognized through the survival of some individuals when they were exposed to those diseases but never became infected. The smallpox vaccine was first discovered by Dr. Edward Jenner. Vaccines function as a precaution before exposure to the illness. They are composed of a physiological fluid and a weakened or dead microbe. Thus, the body recognizes the microbe and produces antibodies or antitoxins to it. The vaccine for each illness is therefore unique. Compound vaccines administered together are used against two or more diseases. Vaccines sustain active immunity, and their effect is long-term.

Properties of vaccines

They should have little or no side effects. Vaccines taken via the oral or nasal passages have the least side effects. Some side effects are high fever and exhaustion.

The age of an individual is important for the success of vaccination, since certain vaccines are more effective and have the least side effects at certain ages. For instance, frequent side effects from typhoid vaccine are observed after the age of 40. When a measles vaccine is administered to an infant less than a year old, eclampsia may result.

Any vaccination should not be given during illness or after surgery.

They have allergic functions.

SCHEDULE OF VACCINATION

Age	Combined injected vaccines
1-2 months	Hepatitis B (two injections, one shortly after birth, another in two months)
2 months	DPT (diphtheria, whooping cough, tetanus) Polio (oral vaccine)
4 months	DPT (diphteria, whooping cough, tetanus)
6 months	DPT (diphteria, whooping cough, tetanus)
12-18 months	Measles, mumps, rubella
15 months	Hepatitis B
18 months	DPT (diphteria, whooping cough, tetanus)
4-6 years	DPT (diphteria, whooping cough, tetanus)
11-12 years	Measles, mumps, rubella

7. Serum

The blood coagulates if it flows out of the blood vessels and is exposed to the air. A straw-colored fluid, known as serum, is separated from the residue. Conversely, if blood is placed into a tube with additives to prevent coagulation, the other constituents of the blood precipitate to the bottom of the tube. The fluid above the precipitated layer is termed blood plasma.

Important Vaccines

Diphtheria vaccine: It can be injected either singly or in combination with whooping cough, tetanus and typhus or paratyphus.

Tetanus vaccine: It can be given at any age. A booster vaccination should be given every ten years. It can be given either singly or in combination with whooping cough, diphtheria and typhus or paratyphus.

Whooping cough vaccine: It is a bacterial vaccine which should be given within the first six months of life.

Typhus-Paratyphus: It is a dead bacterial vaccine which can be given at any age from 3 to 40.

Cholera vaccine: It is also a bacterial vaccine which is given on request to those travelling to high risk areas. Immunity lasts for approximately six months.

Complex vaccines: They are the combination of either 1 or 2 living vaccines or 2-3 dead vaccines.

Immunity is long term:
1. Whooping cough - Diphtheria - Tetanus
2. Typhus - Paratyphus - Tetanus
3. Typhus - Paratyphus - Diphtheria - Tetanus
4. Diphtheria - Tetanus

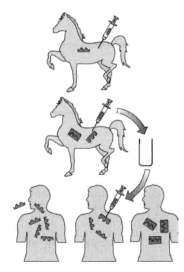

The serum includes only fibrinogen, which is absent in plasma. The serum includes large quantities of protein, antibodies, water and minerals. During illnesses, it is injected into the body to enhance the action of antibodies. It has a short term effect during illnesses.

It can be obtained from an animal that has been infected by a pathogen. This type of serum is called an immune serum. When injected into patients' bodies, it confers temporary immunity against antigens

The serum can be produced in some animals that secrete their antibodies into the blood. The pathogen is injected in increasing doses into a horse, sheep or similar organism. Blood is taken at certain intervals and is processed to remove unwanted substances, such as blood cells and other proteins. The residue is the serum, which can be used in an emergency to support the immune system.

The followings are examples of serums currently available.

Diphtheria antitoxic serum	Tetanus antitoxic serum
Scorpion serum	Snake serum

READ ME

AIDS is caused by a virus called HIV (Human Immunodeficiency Virus). This can damage the body's defense system so that it cannot fight certain infections.

HIV is transmitted in three main ways:

Through unprotected sexual intercourse.

By injected-drug users' sharing equipment, including syringes and

needles From an infected mother to her unborn child

HIV, the virus that causes AIDS, is found in the fluids exchanged during sexual intercourse.

If you inject drugs, and share needles, syringes, mixing utensils or water with an infected person, the virus can easily be passed on to you.

Some people come into contact with HIV through being given infected blood or blood products. Any blood transfusion from these people is a means of HIV infection.

How HIV is NOT passed on:

Everyday contact with someone who has HIV or AIDS is perfectly safe. The virus cannot be passed on through touching, shaking hands or hugging.

You cannot be infected with HIV by touch or sharing objects used by an infected person (cups, cutlery, glasses, food, clothes, towels and toilet seats).

HIV cannot be passed on by sneezing or coughing.

Swimming pools are safe too.

HIV is not known to be passed on through tears or sweat.

You cannot be infected with HIV by mosquitoes or other insects.

Most people who have been infected with the virus HIV can remain healthy for a long time. In fact, many of them may not know they are infected. Some people may have less severe illnesses due to the virus, while others may be quite unwell. From what we know about the condition at present, most people infected with HIV will eventually go on to develop AIDS.

SELF CHECK

CIRCULATORY SYSTEM

A. Key Terms

Aorta	Atrium
Capillary	Diastole
Myocardium	Blood clotting
Hemoglobin	Lymph
Plasma	Rh factor
Serum	Universal donor

B. Review Questions

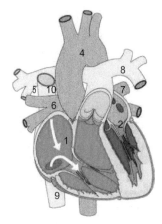

I. Answer the following questions using the figure of the human heart.

1. Label the tricuspid and bicuspid valves.

2. Label the vessels related to the heart.

3. Indicate oxygen concentration in different vessels.

4. Why is the left wall of the heart stronger than the right wall?

5. Explain the importance of exercise to human health.

6. Explain the movement of blood when the atria and ventricles are sequentially contracted.

7. Explain the changes at the atria and ventricles during systole and diastole.

8. Explain the factors involved in movement of blood in vessels.

C. True or False

1. Fibrinogen is a plasma protein needed for blood clotting.

2. Monocytes produce antibodies against unknown antigens.

3. The door-like structures between the ventricles and arteries are called atrio-ventricular valves.

4. The most powerfull part of the heart is the left ventricle.

5. Relaxation of the heart is called systole.

D. Matching

a. Pulmonary artery   ( ) Blood vessel that takes blood away from the heart to the lungs.

b. Sino atrial node   ( ) Small region of neuromuscular tissue that initiates the heartbeat.

c. HLA (Human lymphocyte association) protein   ( ) Protein produced by plasma cells derived from B cells that binds with a specific antigen.

d. Cell mediated immunity   ( ) Protein in plasma membrane that identifies the cell as belonging to a particular individual and acts as a self-antigen.

e. Antibody   ( ) Specific mechanism of defense in which T cells destroy antigen-bearing cells.

Human Systems Help Each Other

Circulatory System

Blood gives nutrients to neurons and remove their wastes. Brain controls heart beat rate and movement of blood in the vessels.

Blood vessels deliver metabolic wastes to kidney, and provide a certain blood pressure to help kidney in filtration. Kidneys filter blood from wastes and maintain pH. Kidneys also produce erythoropoietin which trigger redbone cells production.

Skeletal System

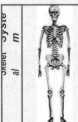

Blood vessels deliver food and oxygen to bones and remove their wastes.

Ribocage protects heart and store Ca^{++}, needed for blood clotting.

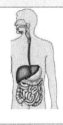

Blood vessels transport nutrients from digestive tract to everywhere within the body. Digestive tract provides Nutrients.

Blood vessels transport hormones to their targets. Heart provides atrial natriuretic hormone.

A group of hormones, including ADH, Aldosteron, ANH regulate blood volume. Adrenalin increases blood pressure.

Blood vessels transport gases to and from lungs. Gas exchange in the lungs helps regulation of blood pH.

Muscular System

Blood vessels supply food and oxygen to the muscles and carry away wastes.

Muscles contraction in heart and vessels providesmovement of blood.

Respiratory System

chapter **7**

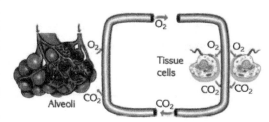

Figure: Oxygen and carbon dioxide are exchanged at alveoli and tissues.

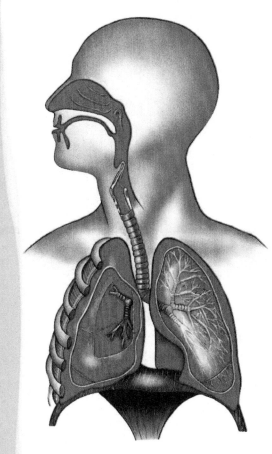

Figure: The respiratory system in humans.

RESPIRATORY SYSTEMS

Respiration is the exchange of gases between an organism and its environment. In small organisms, oxygen can be obtained and carbon dioxide removed by direct contact with the external environment.

In multicellular organisms, the respiratory and circulatory systems cooperate in the transport of gases to the cells. Gas exchange between the respiratory organs, such as the lungs or the gills, and blood is called **external respiration**. The diffusion of gases from the blood to the tissues is termed **internal respiration**. Organisms utilize oxygen in their cellular activities in order to obtain energy. Giant molecules are broken down into their components to extract energy. The resultant energy is used by organisms for metabolic activities such as active transport, protein synthesis, and biochemical and other reactions which require energy in the form of ATP.

These vital activities are performed by the break down of food in oxidation reactions. Oxygen and food react to produce CO_2, water and energy. Carbon dioxide is a poisonous substance which must be removed as quickly as possible by the circulatory system.

1. The Human Respiratory System

In humans and other mammals, oxygen moves into the internal environment of the organism via the mouth or nasal cavity in a process known as breathing. Oxygen is taken in and carbon dioxide is released. The nasal cavity has a rough surface covered by mucosa and contains chemoreceptors, mucosal glands and cilia. The air taken in to the nasal cavity is warmed and moistened and most foreign substances are eliminated before it enters the body cavity.

The air moves initially into the nasal cavity and then into the pharynx and larynx. The larynx is constantly open, except during swallowing, when the **epiglottis** closes the trachea to prevent food from entering it.

The respiratory system consists of two lungs and air passages which transport air to them. The air passages connect directly with the mouth and the nasal cavity. The larynx is located at the back of the mouth, at the top of the trachea. The larynx is a cartilaginous structure which contains the **vocal cords** and protects the trachea. The vocal cords are anchored to the wall of the trachea. The voice is generated by the vibrations of the vocal cords when air flows over them. In speech, the vocal cords move closer to each other, whereas they move away from each other when speech is finished.

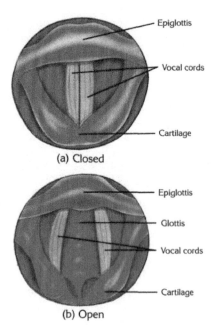

Epiglottis
Vocal cords
Cartilage

(a) Closed

Epiglottis
Glottis
Vocal cords
Cartilage

(b) Open

Figure: The vocal cords function in generation of speech a.
Position of vocal cords during speech. b. Position at rest.

The generation of individual sounds is controlled by the motion of the teeth, tongue and lips. Muscle fibers inside the vocal cords and larynx regulate their tension.

The flow of food into the trachea is prevented by the epiglottis, forcing the food to move into the **esophagus.** The esophagus works in conjunction with the mouth, so preventing the entry of food into the trachea during swallowing. If this reflex mechanism is disrupted, death may result through **asphyxiation** if assistance is not received quickly.

The larynx directs air into the trachea, which is divided into left and right bronchi (major divisions of the trachea that enter the lungs). These bronchi divide into bronchioles (branched tubes that lead from the bronchi to the alveoli) in the lungs.

The left bronchi directs incoming air from the trachea into the bronchioles of the left lung, the same events taking place in the right lung. Each bronchiole ends in a cluster of tiny air sacs, the **alveoli.**

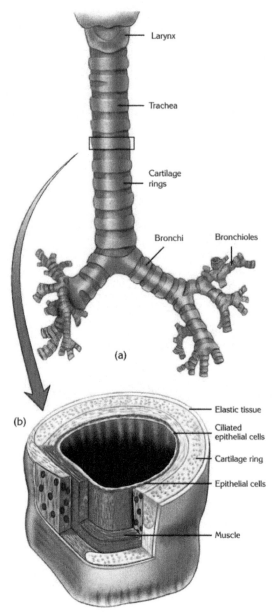

Larynx
Trachea
Cartilage rings
Bronchi
Bronchioles
(a)

(b)
Elastic tissue
Ciliated epithelial cells
Cartilage ring
Epithelial cells
Muscle

Figure:
a. Diagrammatic view of the respiratory tract. It is composed of the larynx, trachea, rings of cartilage, 2 bronchi and many bronchioles.
b. A cut-away section of the trachea showing its components.

Respiratory System

The trachea is composed of three main layers: epithelial cells in the innermost layer, C-shaped cartilaginous rings in the middle layer and a protective connective tissue layer outermost.

The inner layer has unique features, including mucous secreting goblet cells which ensure that the inner surface is constantly wet. Bacteria and foreign substances, such as dust, are trapped by mucous secretions and are ejected from the trachea in the form of phlegm by the ciliated epithelial cells of the inner layer.

The tips of the cartilage rings of the middle layer are interconnected by smooth muscle. The cartilage rings maintain the trachea in an open state under tension. Air moves freely in the trachea due to a difference in pressure between it and the atmosphere. During inhalation, pressure at the trachea is below zero. If there were no rings of cartilage in the trachea, it would close tightly.

There are however, no cartilage rings in the alveoli. A film of lipoprotein lowers their surface tension and prevents them from closing. The lungs collapse in some newborn babies, who lack this film.

a. Lungs

In animals and in humans, the lungs are located in the thorax, which is separated from the body cavity by a **diaphragm**. The right lung has three lobes, whereas the left lung has two. The left lung is slightly smaller than the right lung since the heart is located directly in front of it. Both lungs are surrounded by a thin, double-layered membrane known as the **pleura**. The space between the pleural membranes covering the lung and the pleural membrane lining the thoracic cavity is called the **pleural cavity**. A film of fluid in the pleural cavity provides lubrication between the lungs and the chest cavity. The lungs consist of many similar units, known as alveoli. There are approximately 300 million alveoli in the lungs. The alveoli are arranged regularly to increase the internal surface area of the lungs to

approximately 70-100 m^2.

Alveoli are composed of a single layer of squamous epithelial cells and are surrounded by a network of capillaries. The alveoli are the site of gas exchange in the lungs. Oxygen molecules in air diffuse through the epithelial cells into the capillaries, where they combine with hemoglobin compounds to form oxyhemoglobin. Simultaneously, CO_2 diffuses out of the capillaries into the alveoli and is exhaled. Alveolar cells secrete lipoproteins which coat the internal surface of the alveoli. These substances reduce the surface tension in much the same way as a detergent. Lipoproteins, therefore, prevent water loss from the lungs and provide regular respiration

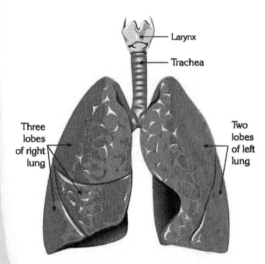

Larynx

Trachea

Three lobes of right lung

Two lobes of left lung

Figure: The appearance of lungs.

At birth, the lungs of infants are slightly hard, and occupy all the unexpanded thorax. The intercostal muscles are free and uncontracted.

As the first breath is taken, the chest cavity expands to a greater extent than the lungs, thus producing a vacuum. Both structures continue to grow rapidly thereafter.

Respiratory Movements

Gas exchange in the lungs is carried out through changes in chest volume, which result in alterations in pressure. The movement of the chest cavity parallels that of the movement of the lungs. The chest consists of the ribcage, the ribs and the intercostal muscles between the ribs. The ribs are connected to the vertebrae at an angle. The contraction of the intercostal muscles results in upward movement of the ribs and expansion of the chest.

The diaphragm, a dome shaped membrane separating the chest cavity from the coelom (abdominal body cavity), flattens during breathing. Thus, the volume of the chest cavity increases.

When the volume increases, the pressure within the chest cavity decreases, becoming less than atmospheric pressure. To equalize pressure inside and outside the body, air rushes in (**inhalation**). The volume of the chest cavity is reduced by relaxation of the diaphragm and the intercostal muscles. At that time, pressure inside the chest cavity becomes greater than atmospheric air pressure, and air is sent out (**exhalation**). The movements of the chest cavity during its expansion and reduction result in changes in the volume of the lungs, since the outer membrane of the lungs is closely connected to the ribcage, aided by fluid in the pleural cavity.

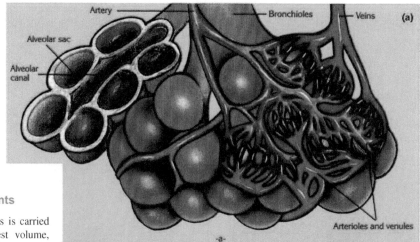

-a-

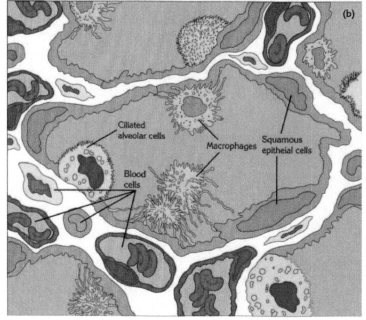

Figure:
a. The bronchioles, detailing the alveoli, pulmonary artery, veins and capillary network.
b. A diagrammatic representation of an alveolus and its components.

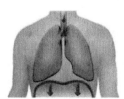

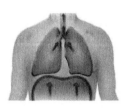

Figure: Respiratory movements
a. During inhalation, the chest cavity expands, the ribs move upward and outward, and the diaphragm flattens. The pressure in the lungs decreases, and air rushes in.
b. During exhalation, the volume of chest cavity is reduced, the ribs move downward and inward, the diaphragm moves upward. The pressure in the lungs increases and deoxygenated air is expelled.

 If the chest cavity of dogs and some other mammals is perforated, both

lungs collapse, shrinking as the membrane surrounding both of them is punctured. Under the same circumstances, only one lung of humans collapses, since each lung is housed separately within a pleural membrane.

When we breathe, the amount of air moved in and out with each breath is called the **tidal volume.** Normally, the tidal volume is about 500 mL, but we can increase the amount inhaled and exhaled by deep breathing. The maximum volume of air that can be moved in and out during a single breath (or the air that can be exhaled after a deep inhalation) is called the **vital capacity.**

Some air, about 1200 mL, stays inside the lungs after exhaling (**residual air**). Inspiration volume can be increased up to 3100 mL; tidal volume is 500 mL; expiratory reserve volume (the air that can be exhaled with force, using the abdominal muscles) is about 1400 mL. Vital capacity is the sum of all these volumes, or about 6 L.

An individual can not respire if the chest cavity is perforated, as air entering through the perforation destroys the vacuum of the lungs. If the chest cavity of some mammals, like dogs, is perforated, both lungs collapse, shrinking as the membrane surrounding them is punctured. Under the same circumstances, only one lung of humans collapses, since each lung is housed separately within a pleural membrane.

c. Oxygen and Carbon Dioxide Transport

The average daily oxygen requirement of a human is approximately 300 liters. This amount may increase 15-20 times depending on physical activity. Carbon dioxide and oxygen are transported by the blood in all animals except insects. The blood of insects is colorless and contains no respiratory pigment, and therefore performs no function in respiration. This explains why the tracheal system in these organisms is involved in gas exchange. One of the unique features of blood is its high oxygen-carrying capacity, which is 6 times greater than that of water. This feature is explained by the presence of respiratory pigments, composed of proteins, and elements such as iron and copper. Animals differ in the types of pigments which give color to their blood.

Hemoglobin: Hemoglobin is found in invertebrate plasma, though it is found in the erythrocytes in all vertebrates. There are normally 280 million hemoglobin molecules in each human erythrocyte. This amount explains the reason for and importance of a high oxygen-carrying capacity. If oxygen was simply dissolved in plasma, blood would have to circulate through the body at a rate of 180 liters per minute to supply enough oxygen to the cells at rest. With hemoglobin, 5 liters per minute is enough. Having hemoglobin is not enough for a perfect mechanism. Keeping them in erythrocytes gives added benefits. If hemoglobin were found only in the plasma, instead of in erythrocytes, the following would be observed: the oxygen-carrying capacity of the blood would be reduced; the density of blood would increase and its viscosity would increase; the heart would have to work harder; and, less oxygen would be transported to the cells.

Hemoglobin is found in dissolved form in invertebrates such as molluscs and segmented worms. This type of hemoglobin is capable of storing oxygen.

Myoglobin is structurally similar to hemoglobin, and is found in skeletal muscle cells. Oxygen binds weakly to it, and during contraction of the muscles it provides an immediate source of oxygen, since as the pressure of oxygen decreases in the muscles, myoglobin releases its oxygen. Thus, myoglobin is a safeguard against oxygen deficiency in muscles. The oxygen-carrying capacity of myoglobin is 25% that of hemoglobin, since hemoglobin may bind with four oxygen molecules whereas myoglobin only binds with one.

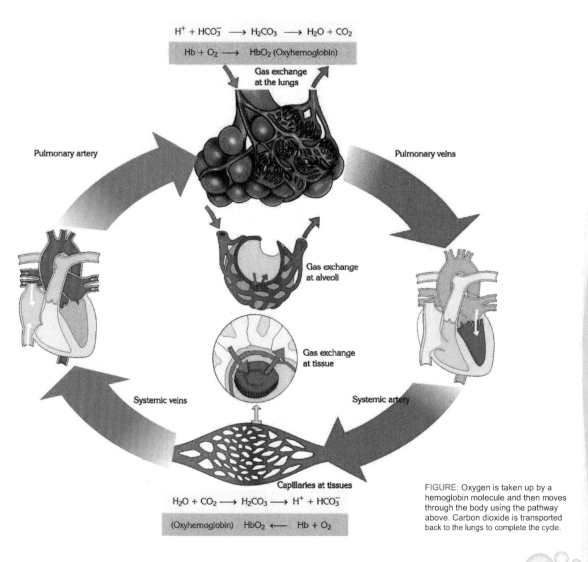

$$H^+ + HCO_3^- \longrightarrow H_2CO_3 \longrightarrow H_2O + CO_2$$

$$Hb + O_2 \longrightarrow HbO_2 \text{ (Oxyhemoglobin)}$$

Gas exchange at the lungs

Pulmonary artery

Pulmonary veins

Gas exchange at alveoli

Gas exchange at tissue

Systemic veins

Systemic artery

Capillaries at tissues

$$H_2O + CO_2 \longrightarrow H_2CO_3 \longrightarrow H^+ + HCO_3^-$$

$$\text{(Oxyhemoglobin) } HbO_2 \longleftarrow Hb + O_2$$

FIGURE: Oxygen is taken up by a hemoglobin molecule and then moves through the body using the pathway above. Carbon dioxide is transported back to the lungs to complete the cycle.

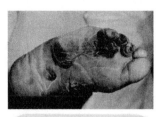

1. Oxygen Transport

Erythrocytes transport 97% of the body's oxygen, and the remaining 3% is transported by the plasma. Air enters the alveoli through the mouth, nose and trachea during inhalation. The oxygen in the air passes through the alveoli into the capillaries surrounding them, and forms **oxyhemoglobin** compounds.

Oxyhemoglobin is a light red molecule and gives this color to the arteries. Molecules are transported to the heart in the pulmonary vein and are pumped to the tissues via the aorta by contraction of the heart. Oxyhemoglobin molecules disassociate into oxygen and hemoglobin in regions of the body where the concentration of oxygen is low. The unloaded oxygen then diffuses into the tissue fluid and into the cells.

The reaction between oxygen and hemoglobin is reversible and is dependent upon oxygen concentration. All hemoglobin molecules are saturated in oxygen-rich regions. Conversely, oxyhemoglobin disassociates into oxygen and hemoglobin in oxygen-poor regions.

At high altitudes, greater numbers of RBC are required to compensate for the reduction in the partial pressure of oxygen. At very high altitudes, oxygen is in short supply due to the low air pressure. As a result, individuals suffer headaches, vomiting, tinnitus (ringing in the ears) and a rapid heart beat. Life is impossible at altitudes higher than 11,000 meters. For this reason, the passenger cabin of a high altitude aircraft is sealed, preventing the interior air from venting. Furthermore, cabin pressure is regulated to approximately 760 mm Hg (1 atm).

2. Carbon Dioxide Transport

The metabolic waste of cellular respiration, CO_2 diffuses from the cells to the tissue fluid and then into the capillaries. About 7% of CO_2 dissolves in the plasma. Another 20% enters the RBC, and combines with hemoglobin to form carboxyhemoglobin. Although it binds to hemoglobin in a different way and at a different site than oxygen, the attachment of a carbon dioxide molecule causes the release of an oxygen from the hemoglobin. Most CO_2 (about 70%) is carried in the form of bicarbonate ions (HCO_3^-). CO_2 reacts with water to form carbonic acid, assisted by the enzyme **carbonic anhydrase**. The carbonic acid then ionizes to bicarbonate (HCO_3^-) and hydrogen ions (H^+). Hydrogen ions are trapped by hemoglobin, whereas HCO_3^- ions diffuse from the RBC to the plasma. Chloride ions diffuse into the red blood cells to replace the negative charges of bicarbonate ions. Carbon dioxide is transported in this way by the veins from the tissues to the heart. It is then transported to the lungs by the pulmonary veins.

The events at the tissues are reversed in the pulmonary capillaries. Bicarbonate (HCO_3^-) ions enter the RBC, reform into carbonic acid, then disassociate into water and carbon dioxide. Carbon dioxide diffuses out from the capillaries to the alveoli and is then pumped out by the constriction of the lungs during exhalation. A RBC releases a Cl^- ion when a HCO_3^- ion is taken up from the blood plasma.

AMOUNT OF GASES IN INHALED AND EXHALED AIR

	% Oxygen	% Carbon dioxide	% Nitrogen
Inhaled (atmospheric) air	21	0.04	79
Exhaled air	15	5.6	79

The bends

Some gases are dissolved in the fluids of the body. At low pressures, these dissolved gases may form bubbles, which may clog the capillaries. This may result if there is a sudden decrease in cabin pressure in an aircraft, and also if a deep sea diver comes to the surface too quickly. Oxygen and CO_2 may be rapidly absorbed, however nitrogen is more likely to clog the capillaries since it remains inert within the blood and can, in such cases, cause possible death or paralysis. This condition can usually be prevented by the gradual release of gases from the body over a long period of time. The individual is placed in a decompression tank where the air pressure is increased, then gradually reduced.

d. Regulation of Respiration

Inhalation and exhalation are under the control of the brain and the spinal cord. The diaphragm and the intercostal muscles are regularly stimulated by the nerves to contract every 4-5 seconds. If nervous stimulation is disrupted or ceases, respiration will also cease. Another factor affecting the mechanism of respiration is carbon dioxide concentration. When the metabolic rate of the body increases, the concentration of carbon dioxide in the blood also increases. As a result the pH value decreases and acidity increases. A low pH value stimulates the respiratory center of the brain, which responds by stimulating the diaphragm and chest. Consequently, carbon dioxide in the blood is removed more rapidly, and the pH of the blood is regulated. Homeostasis is restored.

If the concentration of CO_2 in a closed room increases, however, and the concentration of oxygen is sufficient, the rate of respiration of individuals inside the room will increase. This phenomenon can be explained by the fact that the respiratory center in the brain is sensitive to the blood acidity produced by dissolved carbon dioxide. If the CO_2 concentration is reduced under the same conditions, and the concentration of oxygen is insufficient, the rate of respiration fails to increase, since the respiratory center in the brain is not stimulated. This indicates that carbon dioxide is the main factor regulating the respiration rate. Conversely, the chemoreceptor vessels in the carotid artery can be stimulated by oxygen, and the respiratory center may be activated by the impulse generated, only when the partial pressure of oxygen falls markedly. In addition, receptors in the pharynx and trachea may also affect the rate of respiration. These receptors trigger violent coughing if food enters the trachea. Coughing is the response of the body to prevent accidental intake of food into the trachea.

 The diaphragm and the intercostal muscles are regularly stimulated by the nerves to contract every 4-5 seconds

If the concentration of CO_2 in a closed room increases, and the concentration of oxygen is sufficient, the rate of respiration of individuals inside the room will increase. This phenomenon can be explained by the fact that the respiratory center in the brain is sensitive to the blood acidity produced by dissolved carbon dioxide.

THE REGULATION OF O_2 AND CO_2 CONCENTRATION IN THE BLOOD

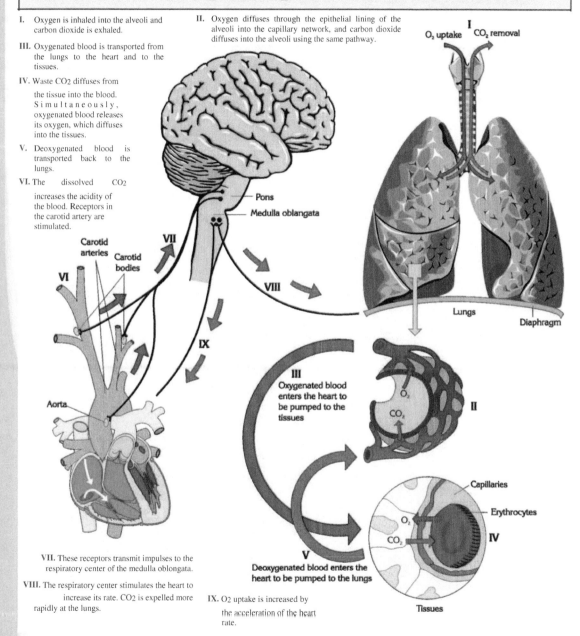

I. Oxygen is inhaled into the alveoli and carbon dioxide is exhaled.

II. Oxygen diffuses through the epithelial lining of the alveoli into the capillary network, and carbon dioxide diffuses into the alveoli using the same pathway.

III. Oxygenated blood is transported from the lungs to the heart and to the tissues.

IV. Waste CO_2 diffuses from the tissue into the blood. Simultaneously, oxygenated blood releases its oxygen, which diffuses into the tissues.

V. Deoxygenated blood is transported back to the lungs.

VI. The dissolved CO_2 increases the acidity of the blood. Receptors in the carotid artery are stimulated.

I O_2 uptake CO_2 removal

Pons

Medulla oblangata

Carotid arteries

Carotid bodies

VII

VI

VIII

IX

Aorta

III
Oxygenated blood enters the heart to be pumped to the tissues

O_2

CO_2

II

Lungs

Diaphragm

Capillaries

Erythrocytes

O_2

CO_2

IV

V
Deoxygenated blood enters the heart to be pumped to the lungs

Tissues

VII. These receptors transmit impulses to the respiratory center of the medulla oblongata.

VIII. The respiratory center stimulates the heart to increase its rate. CO_2 is expelled more rapidly at the lungs.

IX. O_2 uptake is increased by the acceleration of the heart rate.

158

After an accident, the heart and lungs may cease to function, starving the vital organs of the body of oxygen. Unless the normal function of the heart and lungs can be restarted, the casualty will die within a few minutes. It is vital to take prompt action.

Use the 'ABC' plan of resuscitation.

A. Airway : Open the airway by lifting the chin of the casualty. This will tilt the head backwards.

B. Breathing : If the casualty has stopped breathing, use your own exhaled air to inflate his lungs. Pinch the casualty's nose shut. Cover his mouth with yours and exhale.

C. Circulation: The heart of the casualty may have stopped beating. The heart muscle can be stimulated by applying regular pressure to the lower half of the sternum using the palm of one hand.

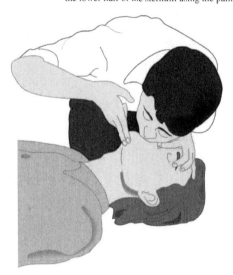
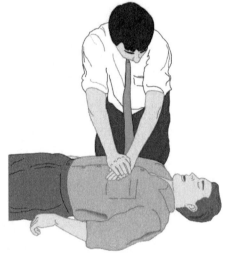

The Abdominal Thrust

If the airway becomes obstructed by food or by any type of hard object entering the trachea instead of the esophagus, the casualty will start to choke to try to remove the obstruction, preventing air from entering the lungs.

If the object does not become dislodged, as a last resort, an abdominal thrust must be attempted. In this procedure, the person assisting should stand facing the back of the casualty, and place a clenched fist, with the thumb facing inwards, in between the navel and the breast bone. Grasping the fist with the other hand, pull them swiftly backwards, squeezing the elbows to compress the lower chest area at the same time. This should remove the obstruction. The procedure can be repeated if necessary.

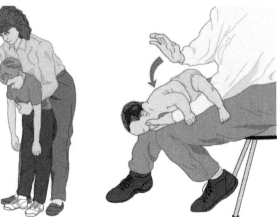

Respiratory System

159

READ ME

Coughing and sneezing are vital reflexes of those who have been infected with the common cold. These mechanisms enable an individual to remove phlegm or mucus from the nose and throat. A cough may result from damage or irritation to the mucosa of the respiratory passageways.

The action of coughing is initiated when the intercostal muscles of the ribs contract rapidly. Since the epiglottis is closed as the muscles contract, the air in the lungs cannot exit easily. The epiglottis then opens suddenly and air leaves the throat noisily.

The type of cough is important in the diagnosis of a respiratory condition. Productive, or phlegm producing, unproductive, persistent or heavy, are all terms by which a physician may categorize a cough.

The mechanism of sneezing is very similar to that of coughing. The events that produce a sneeze are as follows:

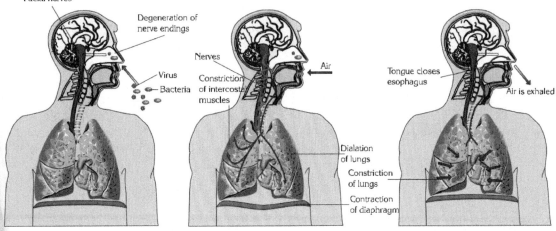

1. **Damage of mucosa layer:** Viruses or bacteria invade the nose and replicate on the mucosal layer, causing irritation which also damages the endings of the trigeminal nerve.

2. **Impulse conduction:** Impulses from the damaged trigeminal nerve endings are transmitted to the cell body of the trigeminal nerve situated on the brain stem. This nerve sends impulses to the intercostal muscles causing them to contract suddenly. It also stimulates the larynx to contract and relax.

3. **Sneezing:** The air displaced by the sudden contraction of the intercostal muscles cannot leave the airways through the mouth and so escapes through the nose. As this air leaves, it is mixed with mucus, so causing the characteristic noisy sneeze.

The whole world is sneezing.

This is how the people of some different countries describe the sound that they make when they sneeze.

German: Hatschi	**Turks: Hapshu**	**Italians: Acci**	**Spanish: Atchi**
Dutch: Atchi	**Iranian: Atchou**	**English: Atishoo**	**American: Ah-choo**

Influenza

Each winter brings epidemics of influenza to some part of the world. Most of those affected will recover completely. The elderly, and those with impaired immune systems are particularly at risk from influenza, which can kill those of this group. The influenza virus presents difficulties in treatment and also in the production of an influenza vaccine. The virus can rapidly change its coat proteins, and many new strains are constantly appearing.

RESPIRATORY SYSTEM

A. Key Terms

Alveoli   Bronchi

Diaphragm   Epiglottis

Larynx   Oxyhemoglobin

Pleura   Residual air

Ribcage   Trachea

Pulmonary circulation   Vital capacity

B. Review Questions

1. Why is gas exchange necessary in living things?

2. Which structures of vertebrates play an active role in respiration?

3. Explain how sound is generated and is converted into speech.

4. Explain the functions of ciliated epithelium in the trachea.

5. Explain the processes that occur in the lungs during respiration.

6. Alveoli are structures found in the lungs of mammals. Explain their functions.

7. What is the role of the nasal cavity in respiration.

8. Explain how gases are carried within the body.

9. What kinds of changes would be observed in mountaineers while climbing from low altitudes to high altitudes?

10. Describe how the rate of respiration is regulated.

C. True or False

1. It is recommended to inhale through the nose rather than the mouth.

2. Oxygen connected to the hemoglobin forms carboxyhemoglobin.

3. The larynx contains the voice box.

4. When the rib muscles and diaphragm countract, exhalation takes place.

5. Amount of air that somebody may exhale after a deep inhalation is called the tidal volume.

D. Matching

a. Exhalation   ( ) Taking air into the lungs.

b. Vital capacity   ( ) Removal of CO_2-rich air from the lungs.

c. Intercostal muscles   ( ) Air sacs of a lung.

d. Alveoli   ( ) Muscles found between the ribs, which help in breathing

e. Inhalation   ( ) Maximum amount of air moved in or out of the body with each breath.

Lungs provide O_2 for neurons and remove their CO_2. Certain centers in brain control respiration.

Lungs excrete CO_2. Lungs and kidneys help each other for regulation of water lost and pH level.

Skeletal System

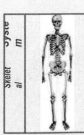

Gas exchange in lungs provides O_2 to the skeletal cells and remove CO_2 which they produce. Rib cage protects the lungs and aids breathing.

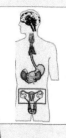

Adrenaline secretion increases the rate of respiration. Lungs provide oxygen to the glands and remove their CO_2.

Gas exchange in lungs decrease CO_2 concentration, which leads pH regulation in the blood. Blood vessels transport gases to and from lungs.

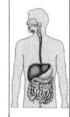

Gas exchange provide O_2 to the organs of digestive system and remove their CO_2.
Pharynx is a common part of both digestive and respiratory systems. Digestive system provides digested food the lungs.

Muscular System

Lungs provide O_2 to the muscles and carry away their CO_2.
Contraction of ribmuscles and diaphragm cause breathing to occur.

BIOLOGY
HUMAN

Digestive System

chapter 8

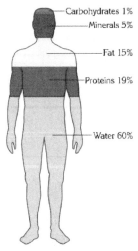

Carbohydrates 1%
Minerals 5%
Fat 15%
Proteins 19%
Water 60%

Figure: The percentage composition of the human body.

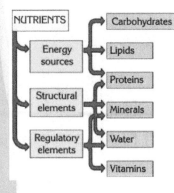

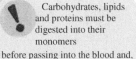

Carbohydrates, lipids and proteins must be **digested into their monomers** before passing into the blood and, consequently, to body cells.

NUTRIENTS AND THE DIGESTIVE SYSTEM

Organisms obtain the energy required for all their metabolic functions, growth and for the repair of their damaged tissues from food.

The energy that food provides is necessary for the continuity of life on earth. Food can be divided into six groups, according to its composition:

carbohydrates	lipids	proteins
vitamins	minerals	water

All of the above nutrients are essential for a balanced diet. A deficiency of any of them may give rise to serious metabolic disorders.

1. Digestive Systems

a. Digestion

In this process, food is ingested from the environment, hydrolyzed into its subunits and absorbed from the gut into the blood. Vitamins, water and minerals may enter the circulatory system without any changes in their composition. Carbohydrates, lipids and proteins, however, require degradation into their monomeric units with the help of enzymes and water, before passage into the blood.

$$\text{Food} + H_2O \xrightarrow{\text{Enzyme}} \text{Monomers}$$

The process of hydrolysis is most frequently carried out by the digestive system. Molecules are split into their component parts with the aid of water. During digestion, large molecules are broken into smaller portions that can cross the plasma membrane easily. This process obviously requires energy in order to carry out the necessary reactions.

The products of digestion are then catabolized within the cell in order to obtain energy for cell repair and maintenance. Some products are stored and consumed when needed. In animals, excess carbohydrates are stored as glycogen, and in plants as starch. Lipids are converted into fat droplets and stored in the adipose tissues of animals. In plants however, they are stored in fruits and seeds (for example, coconuts, sunflower and poppy seeds).

1. Steps in Digestion

Ingestion of food: Food is ingested into the digestive tract through the mouth, aided by the lips, teeth and tongue.

Mechanical digestion: Food is physically ground or chewed into smaller pieces in the mouth by the teeth. A bolus is formed and is swallowed. Peristaltic contractions of smooth muscle in the esophagus, assisted by gravity, transport the food to the stomach. The food is ground further by the muscular activity of the stomach and the small intestine. By means of mechanical digestion, the total surface area of each piece is increased to facilitate effective action of digestive enzymes. In addition to assisting mechanical mixing with digestive enzymes, muscular activity moves food through the digestive tract.

Chemical digestion: Chemical digestion is a series of reactions in which food are hydrolyzed, aided by water and enzymes. Mechanical digestion facilitates chemical digestion by increasing the surface area upon which enzymes can act. The rate of chemical digestion gradually increases from the mouth through the small intestine.

Absorption: Absorption is the final stage of digestion. After the degradation of food in the digestive tract into its monomeric units, the products are absorbed into the blood through the microvilli of the small intestine.

Consequently, they are distributed throughout the whole body via the circulatory system. At the cell they can cross the plasma membrane in order to contribute to protein structure. Some nutrients may also be used as an energy source.

The digestive process is categorized into two parts according to its location:

intracellular digestion

extracellular digestion

a. Intracellular digestion

In unicellular organisms, food is digested within food vacuoles in the cytoplasm. It is taken into the cell by active or passive transport, pinocytosis or phagocytosis. The amount of lysosomes within the cell is directly proportional to the rate of digestion. In organisms that digest intracellularly, they are particularly abundant.

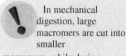
In mechanical digestion, large macromers are cut into smaller macromers, while during chemical digestion, monomers are obtained.

Figure: The process of intracellular digestion. A complex molecule is engulfed by a cell and is broken down by the action of lysosomes into simple usable units.

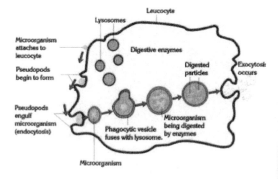

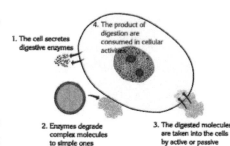

Figure: In extracellular digestion, food material is split into simple molecules before it diffuses into the cell.

In these organisms, a lysosome fuses with a food vacuole to form a digestive vacuole. This association enables enzymatic digestion of food to take place. After digestion is complete, the end products diffuse through the cytoplasm.

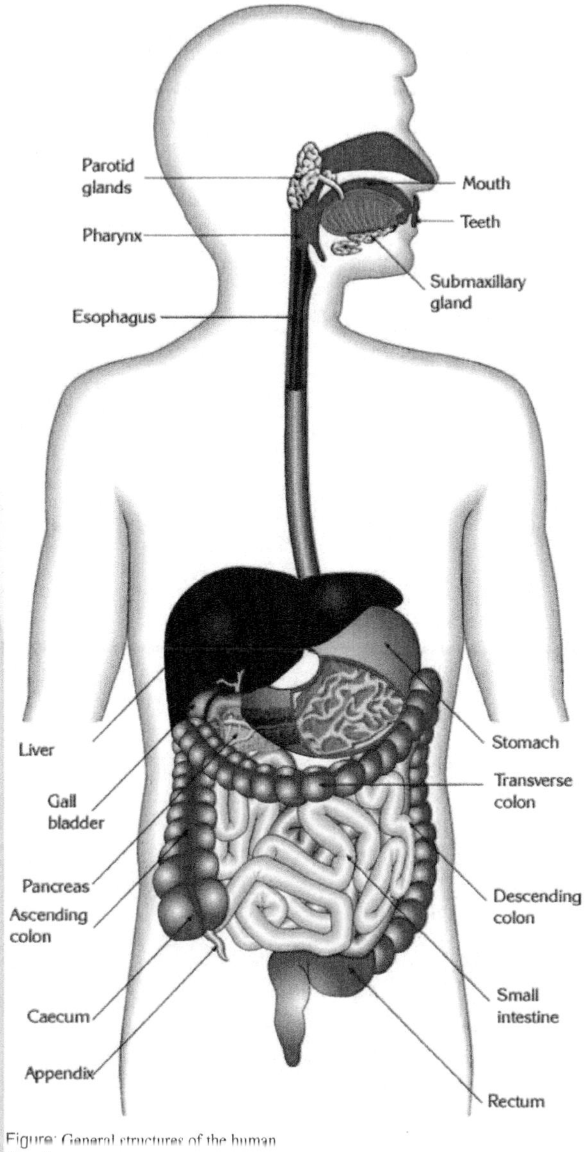

Any indigestible remains are removed from the cell by exocytosis. Intracellular digestion is common among unicellular organisms, such as amoeba, paramecium, and euglena. These organisms cannot digest large molecules and are restricted to those molecules small enough to cross the cell plasma membrane.

b. Extracellular digestion

In this process, digestion of food takes place within an area external to the organism by the secretion of digestive enzymes. These enzymes degrade complex molecules to smaller units so that they may pass through the plasma membrane.

Most organisms using this method of digestion have special digestive organs used in the processing of nutrients.

Extracellular digestion is seen in protista, invertebrates and all vertebrates. It can therefore be concluded that the main purpose of extracellular digestion in organisms is to degrade complex molecules into a usable form.

Digestion in bread molds and saprophytic bacteria:

Protista, such as saprophytes and bread molds, digest extracellularly. These organisms first release digestive enzymes in an extracellular medium in order to break down complex organic molecules. These degraded molecules are then taken in by the cell through different methods of transport.

THE HUMAN DIGESTIVE SYSTEM

The human digestive system consists of a digestive tract in which food is ingested, digested and absorbed. Glands present in the digestive tract secrete substances that assist in digestion.

The digestive tract is comprised of the mouth, pharynx, esophagus, stomach, small and large intestine. The major glands which secrete substances are the salivary glands, alveolar glands of the stomach, glands in the intestine wall, the pancreas and the liver.

Figure: General structures of the human digestive system.

1. Organs of the Digestive System

a. Mouth

The structures found in the mouth that function in digestion are the teeth, salivary glands and the tongue.

b. Teeth

Their function is to grind food into an easily digestible form and to mix it with digestive secretions. The teeth and tongue function cooperatively to form a bolus (chewed food containing saliva) prior to swallowing. The process begins voluntarily and continues as a reflex action. Each tooth is composed of a crown and a root. The root is formed from bone-like **dentine.** The crown, however, is covered with **enamel**, the hardest material in the body. Enamel is formed from calcium, phosphorus and flouride. These minerals should be provided by a balanced diet. If insufficient minerals are ingested, the symptoms of deficiency are seen as dental disorders. Dentine is found directly beneath the enamel, the soft **pulp cavity** in the center of the tooth contains nerves and capillaries. The root is surrounded by **periodontal fibers** and a layer of **cementum** which fixes it firmly to the jawbone.

The dental formula and the formation of teeth

Teeth are formed in special cavities in the jawbone during the last five months of fetal development. Mucosa of the mouth fills each cavity and each newly formed tooth becomes calcified while still in the jawbone. Teeth generally appear 4 to 6 months after birth in the following order: teeth in corresponding positions on the left and the right side of the jawbone appear at the same time. Teeth in the lower jawbone generally appear before those in the upper jawbone. First the central incisors emerge, then the lateral incisors and then the canines. The first molars are next to emerge, and finally the second molars appear. By the end of the third year, the number of teeth has increased to 20. These temporary teeth are known as **milk teeth**. Milk teeth begin to be lost during the seventh year in approximately the same order in which they appeared. Twenty permanent teeth replace them and 8 new teeth appear during this process. The four final molars to emerge do so at the age of 20, and are known as the **wisdom teeth**. An adult has a total of 32 teeth. Dental care, regular brushing and visits to the dentist, can help to maintain healthy teeth.

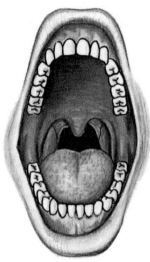

Figure: General view of the mouth.

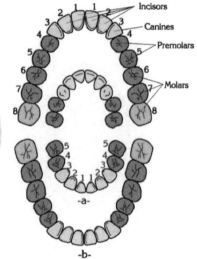

Figure:
a. Milk teeth
b. The position and order of adult teeth.

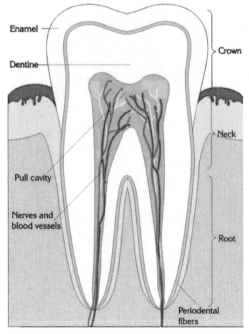

Figure: Structure of a tooth

c. Tongue

The tongue assists the mixing of masticated food with saliva. A bolus of food is formed and swallowed. During chewing, taste buds in the tongue differentiate between bitter, sweet, salty and sour tastes.

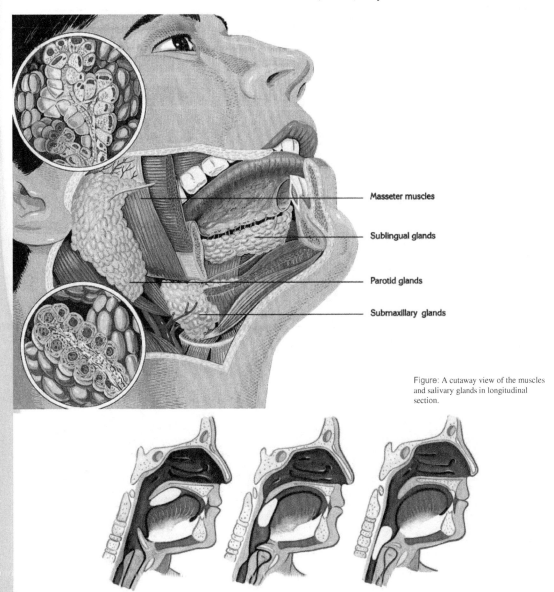

Masseter muscles

Sublingual glands

Parotid glands

Submaxillary glands

Figure: A cutaway view of the muscles and salivary glands in longitudinal section.

d. The Pharynx

The pharynx is a cavity located directly behind the mouth, in front of the esophagus and trachea. It is composed of smooth muscle and membranes, and connects directly with the nasal cavity in its uppermost region and to the esophagus in its lower region. The swallowed food bolus passes from the pharynx to the esophagus. Food is prevented from entering the trachea by the epiglottis. During swallowing, the mouth is closed and the pharynx and larynx rise by muscular contraction. Simultaneously, the trachea is closed by the epiglottis.

e. The Esophagus

The esophagus resembles a pipe, 25 centimeters in length and 2 centimeters in width.

It passes from the pharynx to the stomach. The esophagus is ordinarily collapsed, but it opens and receives the bolus when swallowing occurs. A rhythmic contraction of the digestive tract, called peristalsis, pushes the food along. Peristalsis begins in the esophagus and occurs along the entire length of the digestive tract.

f. The Stomach

The stomach forms the largest portion of the digestive system. It is located on the left side of the body beneath the diaphragm. It is a sac-like structure between the esophagus and the small intestine. When empty and at rest, the stomach is J-shaped. Its dimensions when full are as follows:

25 cm in length   12 cm in width

1250 cm^3 in volume

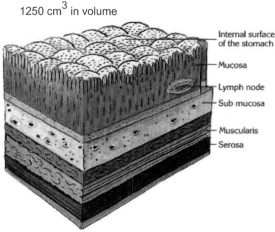

Figure: The structure of the stomach wall.

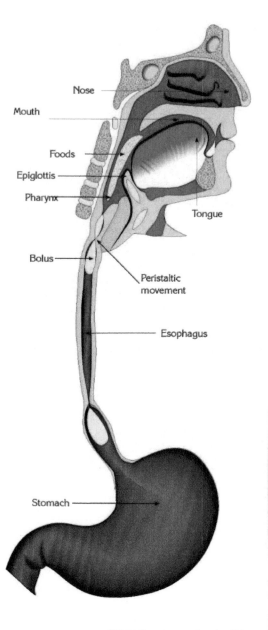

Figure: The structures through which a bolus of food passes from the mouth to the stomach.

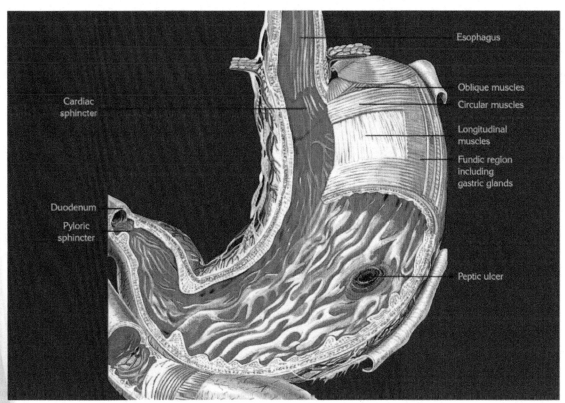

Esophagus

Oblique muscles

Circular muscles

Longitudinal muscles

Fundic region including gastric glands

Peptic ulcer

Cardiac sphincter

Duodenum

Pyloric sphincter

Figure: The muscular structure of the stomach and its associated structures.

 The stomach is the part of the digestive system where mechanical

digestion of all types of food and chemical digestion of proteins occur.

The point where the stomach and the esophagus are connected is termed the **cardiac sphincter**, a valve through which food enters the stomach. The point at which the stomach is connected to the small intestine is termed the **pyloric sphincter**. It regulates the movement of chyme (a soupy liquid) into the duodenum.

Structure: Four main layers constitute the stomach. They are the serosa, muscularis, submucosal layer and mucosal layer.

The serosa: A very thin, outermost layer of the epithelium supported by connective tissue.The rest of the digestive tract has serosa. The serosa secretes fluid that keeps the outer surface of the intestine moist so that the organs of the abdominal cavity slide against one another.

The muscularis: The stomach is composed of three main layers of smooth muscle: circular, longitudinal and oblique. Longitudinal muscles function in vertical motions of the stomach, which shortens when they contract. Similarly, the circular muscles regulate horizontal movements of the stomach. During contraction, the stomach narrows. The oblique muscle layer is located between the cardiac and pyloric regions and regulates the motion in these two areas.

The submucosal layer: It is a layer of loose connective tissue containing a rich network of capillaries and nerves. The submucosal layer is located directly below the mucosal layer.

The mucosal layer: It completely covers the internal surface of the stomach. It is in direct contact with the mucosa of the esophagus at the cardiac region and with the intestinal mucosa at the pyloric region. The mucosal layer contains finger-like, cylindrical gastric glands composed of two types of cells, chief cells and parietal cells.

The chief cells of the gastric glands secrete pepsinogen into the gastric juice, where it is converted to pepsin. The parietal cells secrete hydrochloric acid and intrinsic factors which function in absorption of vitamin B.

The total surface area of the stomach is approximately 500 cm^2, and it contains approximately 5 million glands.

The functions of the stomach:

to store ingested food.

to produce gastric juices to continue the digestion of food.

to mix digestive juices and food by physical movement.

to squirt the partially digested food into the small intestine for digestion to continue and for absorption to occur.

g. Small Intestine

The small intestine comprises the portion of the digestive tract between the stomach and the large intestine, or colon. It is approximately 6 meters in length and 2-4 cm in width. It is folded, however, in order to reduce the area needed to contain it. The small intestine can be subdivided into three main segments.

duodenum   jejunum   ileum

The duodenum is 26 cm long, and 3-4 cm wide. It is the shortest but most active digestive segment of the small intestine. The hepatic duct from the liver and the pancreatic duct both drain into a common duct, known as the **Wirsung channel,** attached to the duodenum.

The jejunum is located directly after the duodenum. It is the longest part of the intestine Its folds fill the lower part of the coelom. The most important function of the jejunum is the absorption of nutrients into the blood.

The ileum forms the final portion of the small intestine where it connects to the large intestine. The small and large intestine are separated by a valve which prevents the back flow of food from the large intestine to the small intestine.

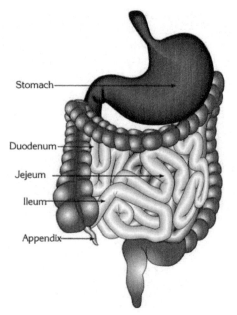

Figure: Showing components of the small intestine.

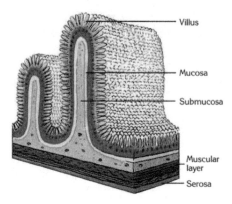

Figure: Showing the arrangement of microvilli.

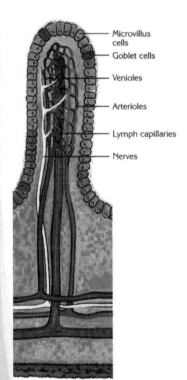

- Microvillus cells
- Goblet cells
- Venioles
- Arterioles
- Lymph capillaries
- Nerves

Figure: A diagrammatic representation of the microvilli on the surface of the villi.

The structure of the small intestine

It is structurally similar to the stomach and consists of four main layers.

The serosa is a continuation of the peritoneum.

The muscularis consists of longitudinal and circular smooth muscle fibres. Food is transmitted by peristaltic movements of these muscles.

The submucosa is structurally identical to that of the stomach.

The mucosa is the continuation of the stomach mucosal layer and covers the internal surface of the intestine. This region is characterised by villi, finger-like projections of the small intestine which absorb nutrients.

Both secretion and absorption occur in the mucosa. This layer contains a large amount of tubular and alveolar glands which release secretions into the small intestine.

The most important mucosal structures are the villi, which are distributed along the small intestine from the Wirsung channel to the colon. Other structures in the mucosa are arterioles, venuoles, lymph vessels and epithelial cells.

h. Large Intestine

The large intestine is located directly after the small intestine. It is a structure 1.5 to 2 meters in length and 6 to 8 cm wide, separated from the small intestine by a valve. The large intestine is subdivided into three segments:

caecum   colon   rectum

The caecum is the blind end of the colon. It is attached vertically to the small intestine. The caceum is approximately 4 to 8 cm long and 5 to 7 mm wide.

The appendix, a projection of the caceum, is reddish in color. This organ may become inflamed and its walls may become hardened, causing severe pain. In such cases, it must be surgically removed.

The colon is the mid-portion of the large intestine between the caecum and rectum.

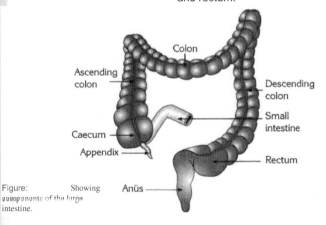

- Colon
- Ascending colon
- Descending colon
- Small intestine
- Caecum
- Appendix
- Rectum
- Anüs

Figure: Showing components of the large intestine.

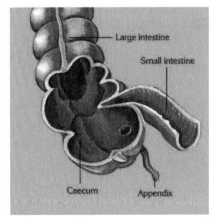

- Large intestine
- Small intestine
- Caecum
- Appendix

The rectum forms the last segment of the large intestine and terminates at the anus.

Structurally, the large intestine is similar to the small intestine. It is made up of the serosa, a muscular layer and mucosa, containing lymphoid structures and glands. There are no villi in the large intestine.

A mixture of indigestible food, still referred to as chyme, is squirted from the small to the large intestine through a valve. It is then moved through the large intestine by peristaltic action of smooth muscles.

2. Digestive Secretions

Digestive secretions originate from the salivary glands, gastric glands, pancreas, gall bladder and small intestine.

a. Salivary glands and their secretions

There are three pairs of alveolar glands, the parotid glands, the sub-mandibular (or maxillary) glands and the sublingual glands. In addition, glands in the mucosa of the mouth produce secretions.

Approximately 1000-1500 ml of saliva is secreted from these glands every day. The secretions of these three glands are mixed within the mouth. These secretions are composed of 3% protein and amylase enzymes, also known as ptyalin. Some individuals develop teeth stones and salivary duct stones. Salivary secretions are regulated by conditional and non-conditional reflexes. In non-conditional, or marital reflexes, food stimulates the nerve endings in the mouth.

The generated impulse stimulates the secretory center of the medulla oblongata in the brain. The impulses generated from this center stimulate secretions from the salivary glands.

Conditional reflexes are observed when the individual exercises, thinks about, sees, or smells food. These actions stimulate the related centers in the brain which initiate saliva secretions.

The parasympathetic nerves of the autonomic nervous system stimulate diluted secretions, whereas sympathetic nerves stimulate undiluted, cohesive secretions.

The function of saliva

> initiation of carbohydrate digestion.
>
> a solvent for detection of taste.
>
> formation of a bolus by mixing with masticated food.
>
> lubrication of the pharynx, so that food may be swallowed easily.
>
> assists in speaking.

Digestion of plant sources is more difficult than that of animal sources.

Thus, plant eaters have longer intestines.

The small intestine is the part of the digestive system where digestion of all food is completed, and absorbtion begins.

Digestive process	Location	Substrate	Enzyme	Enzyme source	Optimum pH	Products	Site of absorption
CARBOHYDRATE DIGESTION	Mouth	Starch	Amylase	Salivary gland	Dextrin Maltose	Dextrin Maltose	Capillary network at the villi
	Duodenum / Small intestine	Starch	Amylase	Pancreas	Alkaline	Maltose	
		Maltose	Maltase	Intestinal glands	Alkaline	glucose+glucose	
		Sucrose	Sucrase	Intestinal glands	Alkaline	Glucose+Fructose	
		Lactose	Lactase	Intestinal glands	Alkaline	Glucose+Galactose	
PROTEIN DIGESTION	Stomach	Protein	Pepsin (in presence of HCL)	Stomach	Acidic	Peptone polypeptide	Capillary network at the villi
		Milk protein	Rennin	Stomach	Acidic	Casein	
	Duodenum / Small intestine	Peptone polypeptide	Trypsin	Pancreas	Alkaline	Peptides, amino acids	
		Peptone polypeptide	Chymotrypsin	Pancreas	Alkaline	Peptides, amino acids	
		Peptides (tripeptide gibi ara bileşikler)	Erepsin	Intestinal gland	Alkaline	Amino acids	
LIPID DIGESTION	Duodenum Small intestine	Lipid	Lipase	Pancreas	Alkaline	Fatty acid, Glycerol	Lymph vessels at the villi
NUCLEIC ACID DIGESTION	Duodenum Small intestine	DNA and RNA	Deoxyribonuclease and ribonuclease	Pancreas	Alkaline	Nucleotide components	Capillary network at the villi

HORMONE	SOURCE	STIMULATING FACTOR	TARGET ORGAN	RESPONSE OF TARGET ORGAN
Gastrin	Gastric mucosa	Entry of food into the stomach	Gastric mucosa	Gastric juice
Secretin	Duodenal mucosa	Acidic pH of the duodenum	Pancreas	Initiation of enzyme secretion
Enterogastrin	Intestinal mucosa	Fatty acids	Stomach	Inhibition of gastric activity
Cholecystokinin	Intestinal mucosa	Acidity of chyme	Gall bladder	Release of bile
Pancreasamin	Duodenal mucosa	Acidity of chyme	Pancreas	Inhibition of pancreatic activity

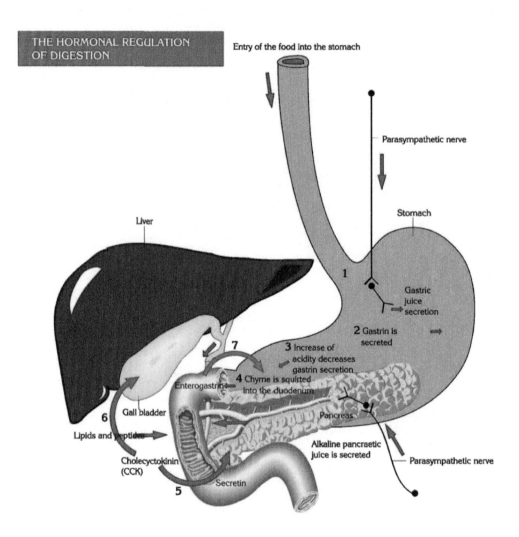

Entry of the food into the stomach

Parasympathetic nerve

Stomach

Liver

1

Gastric juice secretion

2 Gastrin is secreted

3 Increase of acidity decreases gastrin secretion

Enterogastrin

4 Chyme is squirted into the duodenum

Pancreas

Gall bladder

6

Lipids and peptides

Alkaline pancraetic juice is secreted

Parasympathetic nerve

Cholecyctokinin (CCK)

5

Secretin

7

Figure: The secretion of digestive enzymes is controlled by both hormones and by the nervous system.

1. Entry of food into the stomach stimulates secretion of gastric juice by the effect of the vagus nerve.
2. Gastric juice secretion in turn stimulates gastrin secretion, lowering the pH of the stomach.
3. As the pH of the stomach decreases, this inhibits the secretion of gastrin.
4. The presence of chyme in the duodenum stimulates the secretion of enterogastrin, secretin and CCK.
5. The presence of peptides and fats in the duodenum trigger the secretion of pancreatic enzymes, under the influence of CCK.
6. The presence of CCK in turn triggers the release of bile from the gall bladder.
7. The build up of enterogastrin inhibits activity of the stomach.

Digestive System

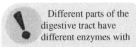

Different parts of the digestive tract have different enzymes with

different optimum pH levels.

b. Gastric Juice

Gastric juice is secreted from the gastric glands located in the mucosa of the stomach, and approximately 300 ml is produced per day.

HCl in gastric juice has the following roles

converts pepsinogen to active pepsin.

stimulates the secretions of the pancreas and intestinal glands.

destroys microbes ingested with food.

facilitates digestion by denaturing proteins.

Mucus, composed of glycoproteins and secreted by the gastric glands, protects the gastric mucosa against peptic digestion and irritation by foreign substances. It forms a protective covering over the internal surface of the stomach. Any damage to the gastric mucosa is repaired by regeneration. Initial abrasion of the stomach is known as a gastrite, which may develop into an ulcer.

The pH value of the stomach is approximately 1 to 2, providing a strong acidic environment for digestion to occur.

Stomach activity is regulated in three phases

Phase 1: The vagus nerve, located at the center of the medulla oblangata, stimulates the gastric secretory cells.

Phase 2: The hormone gastrin stimulates the stomach, causing the release of pepsinogen and HCl. Pepsinogen is then converted to pepsin, which consequently activates the release of more pepsinogen.

Phase 3: The flow of stomach contents into the small intestine stimulates the secretion of enterogastrin, secreted from the duodenum, preventing further gastrin secretion. The lipid and acidic content of blood also activate this reaction.

c. Liver and Bile Secretion

The liver is the largest organ of the body, and weighs approximately 2 kg. Its upper surface is concave and is adjacent to the diaphragm. The lower, convex surface is in close proximity to the stomach, small intestine and the right kidney. The liver is composed of four main lobes. The right lobe is the largest and contains the gall bladder. The liver is supplied with blood by the hepatic artery and the hepatic portal vein. The common bile duct leading from the gall bladder transmits bile salts, which emulsify fats in the duodenum.

The liver is comprised of tissue, the membranes surrounding it and the bile duct. The outer membrane of the liver is known as the peritonea, and connects the liver to the stomach and the diaphragm. A secondary connective tissue located directly beneath the layer of peritonea penetrates into the liver, and is responsible for the presence of lobes. The lobes are 1 to 2 mm in diameter and are pentagonal or hexagonal in shape.

Bile is not an enzyme; it emulsifies fat. The surface area of fat

particles is increased, and lipase can work more effectively.

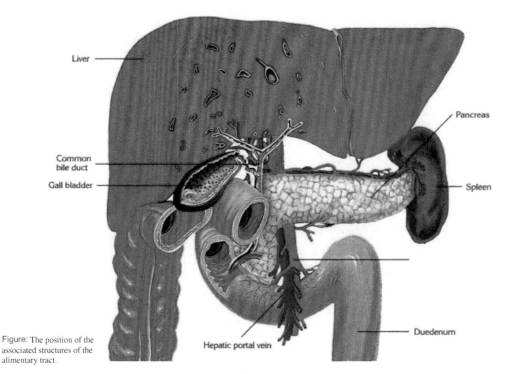

Liver

Pancreas

Common
bile duct

Gall bladder

Spleen

Duedenum

Hepatic portal vein

Figure: The position of the
associated structures of the
alimentary tract.

The lobes are the functional elements of the liver. Each one functions separately and may carry out more than 500 different activities. Each lobe is separated from the others by a connective layer which carries the branches of the hepatic artery and hepatic vein.

These vessels branch out in each of the lobes to form capillaries. The hepatic vein is formed at the point of their reconnection at the center of each lobe, and there are star-like gaps between the lobes. Generally, arterial capillaries are located in the outer portion of the lobe, whereas venal capillaries are found in the inner portion. Kuppfer cells within each lobe eliminate all foreign substances entering the lobe by phagocytosis.

Bile salts are the water soluble metabolic wastes of liver cells. Approximately 1000 ml of bile is secreted from the liver in a single day, and any excess is stored in the gall bladder. Water, sodium chloride and other electrolytes are reabsorbed to concentrate it.

Bile contains cholesterol, fatty acids, plasma electrolytes and bilirubin. Cholesterol accounts for approximately 1 g of the bile, and is insoluble in water. A reduction in bile salts can result in the formation of gall stones by the precipitation of cholesterol. When needed, bile leaves the gallbladder and proceeds to the duodenum via the common bile duct.

The liver is an accessory organ of digestive system. It has more than 500 different functions.

Bile, produced by the liver, aids in fat digestion.

Digestive System

Functions of the liver

Formation and secretion of bile.

Regulation of blood sugar level. The excess glucose in the blood is stored as glycogen in the liver. Glycogen is hydrolyzed and glucose is released into the blood when the level of sugar in the blood decreases.

Formation of fibrinogen and thrombogen, which are active during blood clotting.

Conversion of pro-vitamin A into vitamin A by the enzyme carotinase.

Deposition of Fe, Cu, proteins and vitamins A, D, E and K.

Degradation of old red blood cells by Kuppfer cells.

Elimination of foreign substances, which enter the liver through the blood.

Detoxification of substances.

Synthesis of urine by the reaction of 2 mol CO_2 and ammonia, the waste product of amino acid metabolism.

Heparin, secreted by the liver, prevents clotting of blood within the vessels.

Regulation of body temperature.

The synthesis of red blood cells by reticulo-endothelial cells.

The formation of gall stones

The flow of bile into the small intestine may be blocked by gall stones. Bile is absorbed from the liver and is released into the blood. This condition produces feces that are light in color. Bile pigments also accumulate in the skin, producing a yellowish color. This disorder is known as jaundice, and is not infectious.

The pigments in bile vary from yellow to green or red according to the species of organism. These pigments are formed by the degradation of hemoglobin within the liver. Bile pigments are converted to a brown color by the enzymatic activity of intestinal bacteria, and this coversion of pigment gives feces its color.

Functions of bile

It neutralizes the acidity of chyme entering the duodenum from the stomach, and facilitates the activity of intestinal enzymes.

It emulsifies lipids and facilitates the activity of the enzyme lipase.

It enables the absorption of vitamins A, D, E and K.

It increases absorption by accelerating movement of the villi.

d. Pancreatic secretions

The pancreas is located laterally, directly beneath the stomach, and is connected to the duodenum. It has a leaf-like appearance and is approximately 50 to 80 grams in weight, 16 to 20 cm in length and 2 to 3 cm in width. During

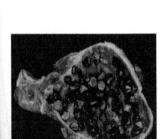

Cholelithiasis

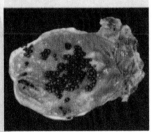

Figure: Gall stones may block secretion of bile.

digestion it is purple in color and is whitish when at rest. The pancreas is composed of lobes, known as acini. Each acinus functions as a small independent unit of the pancreas.

Acinar cells secrete approximately 1.5 - 2.0 liters of pancreatic juices per day. These juices are composed of both organic substances, such as lipase, and inorganic substances, such as water, chloride, bicarbonate and a small amount of phosphate. Furthermore, Na, K, Ca and Mg ions may be found in pancreatic secretions. The other organic molecules are amylase, trypsinogen, chemotrypsin, carboxypeptidase and a small amount of nuclease and lecithinase. These organic molecules act upon different constituents of food.

Lipase acts on fats

Amylase acts on complex carbohydrates

Trypsin, chemotrypsin and carboxypeptidase act on proteins

Pancreatic secretions are therefore capable of digesting the components of most food.

Pancreatic juice has a density equal to plasma, and its pH is between 8.0 and 8.5. Its alkalinity is due to the bicarbonate ions it contains.

Pancreatic secretions are stimulated by the hormone secretin (produced by the duodenal wall). Normally, the inactive form of these pancreatic secretions is found in the mucosa of the duodenum, and is activated by the acidic effects of chyme. Secretine is transported from the cells lining the duodenum to the pancreas by the blood, and stimulates the secretion of pancreatic juice.

Pancreatic secretions are also regulated by CCK (cholecystokinin), produced by the duodenal wall and the vagus nerve.

The islets of Langerhans are located between the acini of the pancreas. The hormones insulin and glucagon, active during the regulation of the level of blood sugar, are secreted here.

e. Small Intestine Secretions

The digestion of food begins at the mouth, continues through the stomach and finishes in the small intestine. Food is completely digested into its monomers by the secretions from the intestinal mucosa, pancreas and liver. The digested nutrients are consequently easily absorbed by intestinal cells.

The secretions of the small intestine are stimulated by the pressure of intestinal contents on its internal surface, partially by the effect of parasympathetic nerves.

3. The Digestion of Food

Proteins, lipids and carbohydrates in food must be hydrolyzed into their monomers before they can be utilized. Teeth, muscles, enzymes and water all play important roles in the digestive process. Carbohydrates, proteins and lipids are gradually altered during digestion. Each different food type is acted on by a different enzyme.

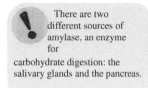

There are two different sources of amylase, an enzyme for carbohydrate digestion: the salivary glands and the pancreas.

Digestive secretions are regulated by nerves and some hormones secreted by the organs of digestive system.

a. Carbohydrate digestion

Carbohydrates in food are in the form of glucose, fructose, sucrose, maltose, lactose, starch, cellulose and glycogen. Cellulose, however, can not be digested by humans or other animals, since no enzyme exists to digest it. Microorganisms in the digestive system are, however, capable of digesting cellulose. The enzymes that digest carbohydrates are called carbohydrases. The sites of carbohydrate digestion are the mouth and small intestine.

$$Polysaccharide + H_2O \xrightarrow{Salivary\ or\ pancreatic\ amylase} Disaccharide$$

$$Maltose + H_2O \xrightarrow{Maltase} Glucose + Glucose \quad Lactose + H_2O$$

$$\xrightarrow{Lactase} Glucose + Galactose \quad Sucrose + H_2O$$

$$\xrightarrow{Sucrase} Glucose + Fructose$$

MOUTH: The digestion of carbohydrates is initiated here. Starch is degraded to maltose (disaccharide) or dextrin by the activity of the enzyme amylase. Dextrin is a part of a starch molecule formed from eight glucose subunits.

STOMACH: No enzyme exists to digest carbohydrates in the stomach. However, foods are mixed under acidic conditions in a process taking approximately 20-30 min. During this period, 40% of starch mixed with amylase is digested. Amylase fails to function when the acidity of the stomach increases.

SMALL INTESTINE: The digestion of carbohydrates is completed in the small intestine. Pancreatic amylase, more effective than salivary amylase, digests all previously undigested carbohydrates to maltose and dextrin. The digestive products of starch and ingested sucrose, lactose and maltose are broken down into their monomers (monosaccharides) by the enzymes secreted by the intestinal mucosa.

Digestion of carbohydrates is completed in the small intestine, and digestive products pass through the villi into the blood.

b. Digestion of Proteins

They are digested in the stomach and small intestine. In the stomach, as explained previously, the hormone gastrin is released into the blood when food enters the stomach. Gastrin functions as a chemical messenger, stimulating the secretion of gastric juice from gastric glands.

Gastric juice includes inactive pepsinogen which, in the presence of HCl, is converted to active pepsin. It is an enzyme which hydrolyzes protein molecules to peptides. Proteins, therefore, are incompletely digested into their monomers in the stomach.

The gastric glands of newborn infants and other mammals secrete the enzyme rennin. This enzyme coagulates milk in order to increase the effect of pepsin.

Stomach mucosa cells → Gastrin

Parietal Cells → HCL

Pepsinogen (From chief cells) → Pepsin

Protein + H_2O → Peptones

Figure: The steps in digestion of protein in small the intestine.

The chyme which enters the small intestine from the stomach is an acidic mixture. It causes the secretion of the hormone secretine from the wall of the duodenum.

This hormone passes into the blood, and at the pancreas it consequently stimulates the secretion of pancreatic enzymes. An effective enzyme, trypsinogen, is secreted, which is inactive upon entering the small intestine.

It is activated by enterokinase, secreted from intestinal glands, and is converted to active trypsin, which hydrolyzes undigested protein molecules. It also digests peptides to amino acids.

Amino acids and the digestive products of protein synthesis pass through to the blood from the small intestine by absorption.

Chyme ⟹ Duodenum cells
(Secretin hormone)

Pancreas ⟹ Pancreatic juice (Tripsinogen, Chymotripsinogen, Enterokinase)

Trypsin

Peptone + Water ⟹ Peptides + Amino acids

Peptidase
Peptide + H_2O ⟹ Amina acids
(Erepsin)

Figure: The process by which proteins are digested

c. Digestion of Lipids

Digestion takes place only in the small intestine. The digestion of lipids is complicated since they are giant, water-insoluble molecules. The bile salts act like detergents to reduce the surface tension of fats. Their action breaks large masses of fat into smaller droplets. The surface area is increased. As a result, lipase, the enzyme which provides chemical digestion of fats, can act more effectively. The digestion of lipids is made considerably more difficult if the gall bladder is removed. For this reason, such people should avoid lipid-rich foods.

Lipid+Lipase+Bile+H_2O →Fatty acids+Glycerol

The enzymes secreted during digestion are hydrolyzed into their component amino acids and reabsorbed after digestion. Protein loss is therefore reduced to a minimum.

4. Absorption

The most important property of the digestive system, absorption, and the digestion of all nutrients is completed within the small intestine. The digested nutrients are then absorbed. Some nutrients, such as vitamins and inorganic substances, pass directly through to the blood without the need for any digestive process.

Absorption from the mouth and stomach is less than that in the small intestine. Some toxins, medicines, ions and the drug cocaine, are absorbed in the mouth. Alcohol, aspirin, some ions, such as K, Na, Cl, Br, and some toxins are absorbed in the stomach.

Absorption occurs mainly within the small intestine, since its internal structure is highly suitable for this purpose. The villi and microvilli increase the absorption

surface of the small intestine to approximately 600 m^2.

Lipid digestion starts and finishes in the small intestine. This

process is done with the help of bile, produced by the liver, and lipase, produced by the pancreas.

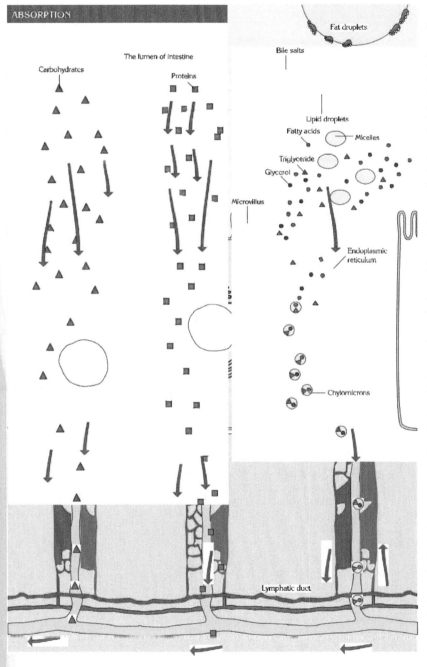

The cells of the microvilli are rich in mitochondria, necessary for active absorption. Each villus contains capillaries and lymph vessels, which transport nutrients both actively and passively from the microvilli of each villus to absorptive cells. The nutrients are then transported from these cells to capillaries and lymph vessels.

The digestive products of carbohydrates, amino acids and vitamins pass from the microvilli to the venules, the smaller branches of veins. The lipid droplets in the small intestine are split into triglycerides by lipase. These triglycerides, emulsified by bile, form micelles, or water-soluble particles.

The hydrophobic fatty acids position themselves at the center of the micelle, whereas the hydrophilic glycerol molecules position themselves on the surface, in direct contact with the water molecules in the environment.

These water soluble micelles are then able to pass into absorptive cells to the endoplasmic reticulum. The micelles are split and the free triglycerides, together with phospholipids, cholesterol and free fatty acids, form chylomicrons, which enter the lymph circulation.

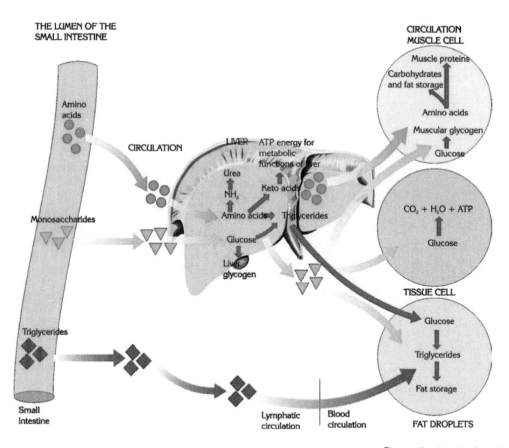

THE LUMEN OF THE
SMALL INTESTINE

Amino
acids

CIRCULATION

Monosaccharides

Triglycerides

Small
Intestine

LIVER ATP energy for
metabolic
functions of liver

Urea

NH$_2$

Amino acids Keto acids

Amino acids Triglycerides

Glucose

Liver
glycogen

Lymphatic   Blood
circulation   circulation

CIRCULATION
MUSCLE CELL

Muscle proteins
Carbohydrates
and fat storage

Amino acids

Muscular glycogen

Glucose

$CO_2 + H_2O + ATP$

Glucose

TISSUE CELL

Glucose

Triglycerides

Fat storage

FAT DROPLETS

Figure: After absorption from the small intestine, organic molecules are distributed to the body tissue.

Lipids facilitate the absorption of lipid-soluble vitamins.

The venules from the villi flow into the hepatic portal vein, which transports all absorbed nutrients from the small intestine to the liver. Blood leaves the liver by the hepatic vein and flows into the inferior vena cava, and finally enters the right atrium of the heart.

Absorption in the Large Intestine

All useful substances are absorbed before the chyme enters the large intestine, and only waste materials, water and a small amount of electrolytes remain. The main function of the large intestine is the absorption of water and electrolytes, and storage of insoluble wastes until their removal.

Furthermore, the symbiotic flora of the large intestine synthesize vitamins B and K. It has been suggested that the only source of vitamin K avaliable for humans is from the symbiotic bacteria of the large intestine.

Anorexia nervosa

Anorexia nervosa is an eating disorder that has a psychological cause. In this condition, the sufferer, usually a teenage girl, is usually of normal weight but perceives herself as severely overweight. As a result, she starves herself, often to the point of damaging her body organs. The exact causes of anorexia are still unknown, but are thought to be linked to low self-esteem and pressure from peers.

Severe cases of anorexia result in hospitalization and forced feeding. The sufferer needs counselling in order to build up her self-esteem.

Cirrhosis of the liver

Cirrhosis is a chronic disease of the liver. It is almost always self-inflicted, resulting from heavy and prolonged drinking, especially in those who have a poor diet. The disease has a high mortality rate. Cirrhosis causes disorders of the circulatory and lymph systems. The cells of the liver become damaged, their functions are disrupted and are finally destroyed.

Those who exhibit symptoms of cirrhosis must stop drinking immediately. For those in more advanced stages of the disease, a liver transplant is a possible option.

Hepatitis

Hepatitis is a disease of the liver and can occur in at least 3 different forms, type A, B and C.

Viral Hepatitis A is the most infectious. It is transmitted through foods that have been insufficiently washed. Seafood is particularly high risk. After an incubation period of up to 45 days, the sufferer develops a headache, loss of appetite, a high fever, dizziness and the whites of the eyes become yellow. Concentration of enzymes produced by the liver increases dramatically.

No medication is prescribed. In most cases, patients are advised to avoid foods containing any fat, and to remain lying down until recovery.

Viral Hepatitis B and C are contracted through blood, and lead to serious complications, such as liver cancer.

Intestinal Obstructions

The clinical and pathologic consequences of obstruction of the gastro-intestinal tract depend upon the level at which the obstruction occurs and whether the obstruction is mechanical or paralytic. Vomiting, with loss of gastric juices and hydrocloric acid, occurs early with high intestinal obstruction, at the level of the pylorus or duodenum.

With obstruction low in the tract, such as the sigmoid region, symptoms are delayed and abdominal distention and acidosis may be leading symptoms. Persistent constipation, vomiting and distention are cardinal signs of intestinal obstruction.

Colicky pains occur with mechanical obstruction. The involved bowel fails to absorb the normal excretion of digestive juices, which amounts to about four liters daily. The loss of these fluids, along with the electrolytes and fixed base, results in dehydration and acidosis or alkalosis, according to the level of obstruction. If the fluid balance and electrolytes are not maintained by intravenous medication, renal function fails and the patient may die in uremia.

Peptic Ulcer (Stomach and Duodenum)

Peptic Ulcer occurs only in the portions of the digestive tract exposed to the action of acid gastric juice. It does not occur in patients with complete achylia. About 10 per cent of the population is afflicted by this ailment at some time. Males predominate over females four to one.

Young and middle aged adults of the lean, worried type are most often affected. Duodenal ulcer occurs with three to four times the frequency of gastric ulcer. Jejunal or marginal ulcers occur following gastro-enterostomy because of the spilling of acid gastric juice directly into the jejunum.

Duodenal ulcer is rarely associated with cancer; the high acid secretion seems to protect, but gastric ulcer is complicated by gastric carcinoma in seven per cent of the cases. Peptic ulcer is caused by digestion of the mucosa and muscularis by acid gastric juice that is (a) excessive in quantity or acidity; (b) improperly neutralized by bile or pancreatic juice; or (c) is improperly neutralized by inadequate gastric mucus secretion.

The modern theory is that psychic or physical stimulation of vagus impulses causes increased gastric motility, hypersecretion, hypercholorhydria and vasospasm, leading to ischemia, infarctions and erosions of the mucosa. Erosions or ulcerations of the gastric mucosa can be produced experimentally by prolonged injections of histamine, which stimulates acid secretion, or hydrochloric acid itself. The pain of peptic ulcer is dependent upon the presence of acid chyme.

Peristalsis and muscle tension induce pain only if the acid is present. The daily rhythm of the ulcer cycle, and its relief by soda or food, is dependent upon alterations in free gastric acidity. Both the pain and the lesion of peptic ulcer result from the action of acid gastric juice on an inadequately protected tissue.

Acute Gastritis (Gastro-enteritis, Food Poisoning)

This is characterized by sudden nausea and vomiting, accompanied by abdominal cramps and followed by diarrhea. It is caused by eating food contaminated with bacterial toxins or organisms of the staphylococcic or salmonella group. Usually a group of persons who have eaten the contaminated food are affected.

On examination, the abdomen is profusely sore, and on auscultation hyperperistalsis may be heard. Nausea and vomiting usually subside within a few hours, but abdominal discomfort and diarrhea persist for several days. The wall of the stomach is covered by a thick mucus secretion, the mucosa is reddened and contains petechial hemorrhages. These signs on gastroscopic examination outlast the subjective symptoms of the patient. Apparently the small intestine is similarly affected. Fatalities are practically unknown except in infants.

Acute Appendicitis

Appendicitis usually is caused by colon bacillus, and anaerobic staphylococci or streptococci and occasionally by spirochetal organisms or *Actinomyces bovis*. The precipitating factor is obstruction by angulation of the appendix, blockage with worms. fecaliths, foreign bodies of undigested food, inflamed lymphoid tissue from other infections, etc.

The organisms contained in the fecal material in the lumen of the appendix cause focal ulceration when infection penetrates. This is characterized by infiltration by polymorphonuclear leukocytes of the submucosa, mucosa, muscularis and finally the serosa. This diffuse inflammation of all coats of the appendix, which usually affects the distal two-thirds of the organ, is known as diffuse acute appendicitis.

The organ may show infarction (gangrene) and rupture. The infection is then followed by localized abscess, peritonitis or pylephlebitis with liver abscess (if the portal vein is the site of metastasizing thrombophlebitis).

Chronic Appendicitis

There are several varieties of inflammatory lesions of the appendix which are grouped under the term chronic or subacute. In chronic appendicitis, which results from repeated acute attacks, the organ is larger than normal, is firm and indurated, and, when sectioned, its walls are thickened. There is evidence of mucosal ulceration and fibrosis under the microscope.

In chronic obliterative appendicitis, the wall and the lumen of the appendix are largely replaced by fibrous tissue which encloses small amounts of fat. It is believed that some cases of obliterative appendicitis represent abnormalities of development and that the absence of the appendiceal lumen represents a form of congenital atresia.

The third type of chronic appendicitis is that characterized by hypertrophy of the lymphoid tissue. This has been termed chronic lymphoid appendicitis. In this form, the appendix is swollen, pale and has a thickened mucosa and submucosa. The outstanding feature is the marked hyperplasia of lymphoid follicles and lymphoid tissue.

Patients with this disease have repeated bouts of lower abdominal cramps and pain, but lack the acute symptoms of muscle spasm and rigidity.

Constipation

Normally, an individual empties his bowels regularly once or twice a day. Constipation results when material moves through the colon too slowly and indigestible matter for excretion is stored in the rectum but is not excreted. The longer the fecal material remains in the colon, the greater the reabsorption of water, making defecation more and more difficult.

The straining involved in expelling waste may lead to hemorrhoids.

The causes of constipation may be due to a low-fiber diet and stress.

Sufferers should avoid taking laxatives. Instead they should try to gradually increase the amount of fiber in their diet and to eat regularly.

NUTRITION

Nutrition is the utilization of ingested substances by a healthy individual for growth, development, reproduction and survival. Research has shown that there are at least forty different types of nutrients necessary for the continuation of life functions. These nutrients are classified as carbohydrates, lipids, proteins, vitamins, minerals and water according to their chemical structure and their different roles.

1. Carbohydrates

The principle function of carbohydrates is the production of energy. Carbohydrates contribute to the structures of ATP, NAD, vitamins and the nucleic acids DNA and RNA. Indigestible carbohydrates, such as cellulose, facilitate the passage of material through the large intestine. Furthermore, in this region *E. coli* bacteria digest some of the cellulose and synthesize vitamin-K.

The Storage of Excess Carbohydrates in Tissues

Carbohydrate requirements: If the daily amount of carbohydrate fails to be supplied, energy is obtained from the catabolism of proteins and lipids. The blood pH decreases and homeostasis is lost. Urine accumulates in the blood and causes disorders such as uremia.

 There are at least forty different types of nutrients necessary for the continuation of life functions. These nutrients are classified as carbohydrates, lipids, proteins, vitamins, minerals and water, according to their chemical structure and their different roles.

Carbohydrates are abundant in cereals and their products, vegetables, fruits and legumes. Carbohydrates are the cheapest and most available source of energy. Excess carbohydrate in the body is converted into lipid, and is stored as adipose tissue, resulting in obesity. People who are obese should reduce their intake of carbohydrate. A diet high in soluble carbohydrates results in dental caries.

Glucose is the main source of energy for many cells of the body, including neurons.

Excess glucose is stored in the liver and muscles in the form of glycogen.

Carbohydrate Metabolism

Nutrients containing starch and sugar are catabolized into glucose in the digestive system. Consequently, glucose units are absorbed into the blood from the small intestine. The amount of glucose in blood remains constant at 100-120 mg. Any excess glucose is stored as glycogen in the muscles and liver, and glycogen is converted into glucose units if the level of glucose in the blood decreases. Over activity of muscular tissue, in the case of prolonged exercise, produces lactic acid. This is transmitted from the tissues to the liver and is converted into glycogen, an insoluble storage form of glucose. High quantities of glycogen are stored in the muscles of athletes as an immediate source of energy.

2. Lipids

These giant molecules yield the most energy in comparison to other molecules. Excess lipid consumed is also stored in adipose tissue. In the event of a lack of blood sugar, these stored lipids are hydrolyzed. Lipids play important roles, as they provide for the uptake of lipid soluble vitamins into the body, a friction-free surface for articulation of the joints, and protection of organs by the formation of a thick covering. The amount of lipid molecules varies in animal and plant food. Fatty vegetables include olives, corn and nuts, such as peanuts and walnuts. Lipids are present in fatty tissue, around organs and under the skin. Animal products, such as milk and eggs, contain lipid molecules. Lipids, in fact, form the bulk of milk. In eggs, lipid is concentrated in the yellow yolk.

3. Proteins

They are the basic structural elements of the body. The development of an organism from a zygote and the formation of its organ systems are entirely dependent on proteins. Furthermore, proteins form part of the molecule hemoglobin, which transports O_2 and CO_2 in RBC, and in antibodies, which defend the body against microorganisms.

Additionally, proteins sustain the osmotic balance between the intracellular and extracellular environment. Homeostasis is lost when the protein balance of the blood in the tissues changes. Biological catalysts, known as enzymes, are proteinaceous in structure. They also participate in the structure of hormones, which are regulatory molecules in the body. In emergencies, proteins may be used as an energy source. Proteins are found in all plant and animal tissues in different quantities (figure-3.43). However, animal foods contain more protein as compared to plant tissues.

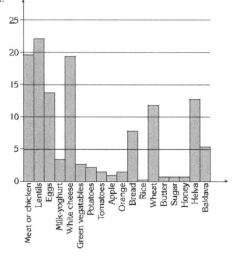

Figure: The proportion of protein per 100g of common foods.

As has already been stated, proteins contain 20 different amino acids which are structurally and functionally different subunits. Some of them are synthesised within the body and some are imported, since they are unable to be synthesized in the body. The compulsory (vital) amino acids that can not be synthesized in the body are leucine, isoleucine, lysine, tryptophan, threonine, methionine, valine and phenylalanine.

Foods vary in their quantity of vital amino acids. Unqualified proteins contain fewer vital amino acids, and are consequently digested with difficulty. The qualified proteins contain adequate amounts of vital amino acids and are easily digested. Generally, animal proteins are qualified but plant proteins are nonqualified.

4. Vitamins

Vitamins are also required for a balanced diet, in addition to carbohydrates, lipids and proteins. A small amount of vitamins is ingested in food and plays important roles in regulation of the metabolism of the body.

Most vitamins function as coenzymes in digestion and participate in metabolic reactions. Many enzymes, therefore, require vitamins for their normal function. With the exception of vitamin D, vitamins can not be synthesized in the body, and must be directly ingested. If there is a deficiency of a particular vitamin, symptoms rapidly follow. Conversely, excess intake of fat soluble vitamins has a toxic effect.

Vitamins were first discovered in 1890 when the disease beriberi was found to be due to a lack of vitamin B. After that initial discovery, different types of vitamins have since been identified, but most of them and their effects are still unknown. In recent years, some synthetic vitamins have been developed that are more powerful than their natural counterparts. Synthetic vitamins are functionally and structurally similar to natural vitamins.

The main source of vitamins is plants. However, animal tissues, especially liver, contain a rich supply of vitamins. In the first and second trimester of pregnancy, during the developmental phase of the fetus, mothers should be encouraged to eat foods high in vitamins, such as liver.

Vitamins, however, are unstable compounds. Vitamins A, B, E, and K are all affected by light. Vitamins A, C, D and E are all affected by oxygen. Vitamin C and E are also affected by exposure to iron and copper. Furthermore, vitamin structure denaturates at high temperature. Overheating of food, therefore, may cause destruction of vitamins.

Functions of vitamins

to give the body resistance to infection.

to prevent against bleeding and blood deficiency.

to assist in formation, development and rigidity of bone tissue.

to regulate growth, development and reproduction.

to provide a regular program of nutrition.

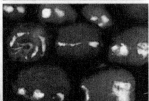

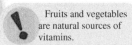
Fruits and vegetables are natural sources of vitamins.

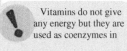
Vitamins do not give any energy but they are used as coenzymes in

metabolic activities.

5. Minerals

Inorganic molecules, or minerals, are as important for body functions as organic molecules. They are required for health, continuity of metabolism and in the formation of bones and teeth. They are divided in to two main groups:

Essential minerals (calcium, phosphorus, sodium, potassium)

Nonessential but recommended minerals (magnesium, iron, copper, zinc and other trace elements)

6. Water

It is the prerequisite for life. Humans can survive for many days without food. Without water, however, death occurs within a few days. Water constitutes 60-70% of the body of an adult. *During growth, from infant to adult, the quantity of water decreases and the quantity of fat increases.*

Functions of water

Absorption, transport and digestion of food

Excretion of metabolic wastes

Regulation of body temperature

As a shock absorber, providing protection against impact and

friction In the absence of water, enzymes fail to function

ENERGY

Half of the chemical energy stored in food is produced by cell respiration and is consumed in the synthesis of ATP. The other half is released as heat during these reactions. The body needs a continuous supply of energy, as ATP is constantly consumed.

The capacity of energy in food

This is measured by its complete combustion in a calorimeter. The units of energy released by food during combustion are measured in calories or kilocalories (kcal).

A Kcal is the amount of energy needed to increase the temperature of 1 g of water by 1°C (from 15°C to 16°C)

1 g carbohydrate	-	4.1 kcal
1 g lipid	-	9.3. kcal
1 g protein	-	5.3 kcal

These carbohydrates and lipids are oxidized to CO_2 and H_2O inside the body, with the release of the same products as if they were in a calorimeter. Proteins, however, are incompletely oxidized and the end products are excreted as urine or as other nitrogenous wastes.

Lipids give at least two times more energy than carbohydrates and proteins. However, their digestion is difficult and gives rise to some dangerous end products. That is why the body uses carbohydrates, but not lipids, as the primary source of energy.

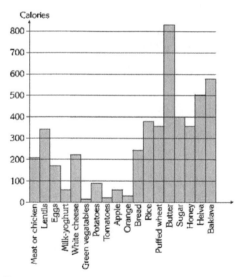

Figure: The calorific value of common foods. This value is dependent on the type of organic molecule present and its proportion. Food containing a large amount of fat has a high calorific value.

Daily energy requirement of an organism

Basal metabolism, the energy requirements of an individual at rest, is determined at 18-20°C (at room temperature).

The consumption of oxygen indicates the amount of energy produced in the body. The amount of energy in the body is directly proportional to the rate of metabolism.

The basal metabolism is approximately 1700 kcal for males and 1600 kcal for females, who both have a body surface area of 1.8 m^2.

The daily energy requirements depend on activity, food ingested, the temperature of the environment and the body, body surface, age and sex.

For instance, the energy requirement of an infant is higher than that of an adult since the body surface per kilo is higher in infants. Thus, the basal metabolism is faster than in adults. Secondly, infants require more energy for development.

Daily Food Requirements For a Balanced Diet

The recommended daily intake is 500 g of carbohydrate, 70 g of lipid and 70 g of protein.

The energy requirements of organisms with heavy bodies are obviously greater than organisms with light bodies.

During the developmental phase of organisms, calorie consumption is high due to a high metabolic rate.

In the case of disease, cellular activity increases, especially in the recovery period, requiring a greater number of calories.

Overactivity of the thyroid gland affects the basal metabolism and calorie consumption increases.

The basal metabolism of males is higher than that of females.

Energy consumption depends on the type of work undertaken. The table above outlines the energy requirements of an adult for certain activities.

The body weight will remain constant only when the energy consumed by the body is replaced by ingestion of nutrients from the environment. In case of an imbalance due to more food ingested than the body needs, this excess energy is stored as lipid. These lipids are consumed when the energy available from ingested food is less than the energy consumed in normal metabolic processes.

Energy requirements are also related to physical activities. A normal resting male requires 2234 kcal of energy whereas 1770 kcal of energy is required for a resting female. However, a male manual worker consumes 3657 kcal while a female manual worker requires 2876 kcal of energy.

People should adapt their intake of nutrients according to their energy consumption. For instance, a resting individual has little need for a high energy lipid diet.

In humans, growth is extremely rapid up to 18 years of age, decelerates between 18 and 25, and ceases completely after 25. Over this age energy needs are reduced. Unless energy intake is regulated, obesity can not be avoided.

The normal body weight can be calculated as follows.

B.M.I (Body-Mass index): It is calculated as 21 for females and 22 for males, but varies according to the individual. The minimum is 19-20, and 24-25 is the maximum value.

B.M.I. = $\dfrac{\text{Weight (kg)}}{[\text{Height (m)}]^2}$

Example: Calculate the appropriate weight for a man and a woman who are both 1.60 m in height.

for a man 22 = $\dfrac{\text{weight (x)}}{(1.60)2}$ = 56.32

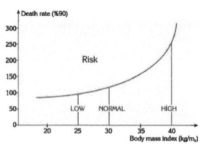

Figure: Death due to obesity carries high risk in those between the age of 35 and 55. The body mass index indicates those who are most at risk.

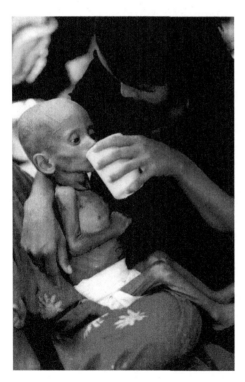

The ideal weight of a female is approximately 53.76 kg. An acceptable range extends from 49 to 61 kg. Below 49 kg, the individual should take steps to increase her body weight by consuming more high energy food. Above 61 kg, she should start to diet.

An individual who intends to lose weight should continue to eat three meals per day. These meals should be mainly vegetables, and sugary or fatty foods should be avoided.

A common mistake when dieting is to skip meals and eat more in the evening. The body can not utilize the extra energy, as individuals are generally less active in the evening. The excess energy is stored as fat in adipose tissue.

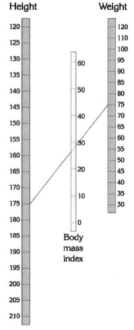

Figure: The correlation between body height and weight using the body mass index.

Digestive System

The causes of obesity

Eating until the stomach is full and the individual feels bloated, in addition to the habit of eating 3 heavy meals a day since childhood, is clearly inviting obesity. Such compulsive eating habits make therapy and treatment more difficult. Since satiety of hunger varies between individuals, obesity during childhood is difficult to treat and explains why mothers should be discouraged from overfeeding their children.

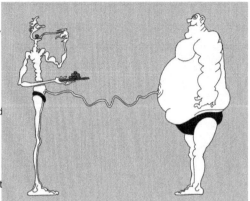

There is, in reality, very little that is known about the causes of obesity. What is known is that the effects of the factors that cause obesity can be lessened by increasing the amount of energy expended or decreasing the amount of energy taken in in food, or both. The basic reasons for obesity that are known to date are:

1. Environmental factors:

Research conducted on twins living in different environments has provided valuable results. A change of lifestyle for example, may result in alterations in body weight. An individual is more likely to gain weight if he has a car or if he increases the quantity of his food intake while at work, for example.

A change in the eating environment may result in a change in the type of nutrition. The type of oil or fat used in meal preparation, the amount of carbohydrate in food, as well as the quantity of the meal, may vary according to the environment. "Fast food" habits in big cities and eating in cafes also cause obesity. The place of eating clearly influences the amount of energy ingested in food.

2. Genetic tendency:

Recent research supports that 75% of those who are obese have either an obese father or mother; 25% have both an obese mother and father. If only one parent in a family is obese, the probability of their children also being obese is 40%. If, however, both parents are obese, there is an 80% probability that the children will be obese. Can these statistics be related to the environment being the same for both parents and children?

It has been found that the energy expended by babies of obese mothers is lower than babies of thin mothers.

It is also clear that biological parents are more effective in the determination of body weight of their children than parents who adopt a child.

If genetic factors are involved, obesity becomes more dangerous. In the case of normal genetic factors, the onset of obesity is during childhood or the teenage years. If, however, genetic factors are only slightly involved, obesity appears after the age of 20.

3. Hormonal factors:

If obesity is due to some hormonal disease, the condition may improve.

4. Neurological factors:

This factor is only rarely effective. In the case of damage to the hypothalamus, obesity may result.

5. Psychological factors:

Psychological factors are extremely effective in both obesity and its treatment and are one of the main factors of obesity.

The emotional state of an individual can also affect his appetite and subsequent food intake. Some individuals increase their quantity of food intake when they are angry, others show the reverse tendency.

A. Key Terms

Amylase   Appendix

Bile   Duodenum

Esophagus   Gall bladder

Lipase   Pancreas

Salivary gland   Sphincter

Villus   Vitamin

B. Review Questions

1. Which of the following substances can be directly injected into blood vessels to feed the body: galactose, vitamin C, peptone, lactose, fatty acids?

2. 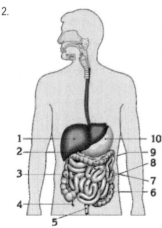 Label the components of the digestive system shown in the figure.

3.
Maltose+Water →Maltase ....

Lactose+Water →Lactase ....

Sucrose+Water →Sucrase ....

Fill in the blanks with monosaccharides which are obtained by the digestion of the above disaccharides.

4. List the organic molecules found in the human body.

5. Which organic molecule is not involved in the production of energy?

6. Where does digestion stop in the human body?

7. In which parts of the alimentary tract is food digested mechanically?

8. Explain the stages of chemical digestion.

9. List the vessels which transport nutrients from the small intestine to the heart.

C. True or False

1. The mouth has no role in chemical digestion of carbohydrates.

2. The liver is not an organ of the digestive system, but its product, bile, helps fat digestion.

3. Pepsinogen and HCl are products of the pancreas.

4. The large intestine is responsible for water and mineral reabsorption

5. Vitamins can not give any energy to an organism, but are needed for metabolism.

D. Matching

a. Colon   ( ) Lymphatic vessel in a villus of the intestinal wall.

b. Defecation ( ) Large intestine that extends from the caecum to the rectum.

c. Lacteal   ( ) Discharge of feces from the rectum through the anus.

d. Lipase   ( ) Fat digesting enzyme secreted by the pancreas.

e. Pepsin   ( ) Protein digesting enzyme secreted by the gastric glands of the stomach.

Digestive system provides nutrients to neurons and glial cells.

Brain controls muscles of the digestive tract.

Elimination of metabolic wastes is done by kidneys.

Digestive tract supplies nutrients to the kidneys.

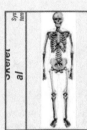

Digestive tract provides minerals including Ca^{++} for bone growth and

Bones provide support and protection for the organs of digestive system.

Some organs of digestive system; stomach, small intestine, produce hormons. Hormones control secretion of digestive glands.

Digestive tract provides nutrients which are delivered to the body by blood vessels.

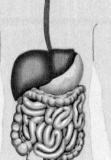

Pharynx and mouth are common structures for both respiration and digestion. Digestive tract provides nutrients to the lungs.

Muscular System

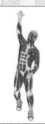

Digestive tract provides glucose and other nutrients for muscle activity.

Peristalsis is provided by smooth muscles.

Skeletal muscles protect the organs of digestive system.

Excretory System

chapter **9**

During metabolic activities, organic molecules are utilized to produce energy. However, some metabolic substances are also produced. By means of the excretory system these harmful substances are eliminated from the body.

EXCRETORY SYSTEM

In the preceeding chapters, the digestion of food and the uptake of oxygen has been illustrated. Digested food and oxygen are transported to the cells by the circulatory system. The cells utilize these molecules in their metabolism. In this chapter, the method of expulsion of metabolic wastes excreted from the body and the structures involved in these processes will be discussed. Excretion rids the body of metabolic wastes, which come from the breakdown of substances. The functions of the excretory system can be summarized as follows: filtration and excretion from the blood of toxic wastes produced by the metabolic reactions of cells; the maintenance of homeostasis by the balance of water and the ionic content of the blood and tissue fluid; the mainteinance of the normal functions of cells; and, the regulation of blood content

1. Excretory Substances

The metabolic wastes of cells are water, carbon dioxide and nitrogenous compounds.

a. Water and Carbon dioxide (H_2O, CO_2)

They are generated during the catabolism of carbohydrates and lipids.

Additionally, deaminated amino acids release H_2O and CO_2 during cell respiration. Water and carbon dioxide are excreted by the lungs, kidneys and by sweating.

b. Ammonia (NH_3)

Amino acids must be deaminated in order for catabolization to take place. As a result, ammonia is formed from the amine groups. It is highly toxic and requires considerable dilution in water for it to be excreted safely. It is thus the excretory substance of freshwater organisms, such as paramecium, sponges, coelenterates, flatworms and freshwater fish.

c. Urea

It is the nitrogenous waste of organisms living in a watery environment and is less toxic than ammonia. Urea dissolves in water and is excreted by the kidneys. Urea is formed by the reaction of 2 mol ammonia with 1 mol carbon dioxide. These reactions form part of the ornithine cycle

$$2NH_3 + CO_2 \rightarrow CO(NH_2)_2 + H_2O$$

d. Uric acid

A small amount of water is utilized during excretion of uric acid due to its inability to dissolve in water. Uric acid is ideal as the excretory product of organisms that need to conserve water.

Figure: The formation of soluble nitrogenous waste is the product of deamination of proteins during cell metabolism. In the first stage, keto acids and ammonia are produced. The keto acids can be transferred for use in the Krebs cycle.

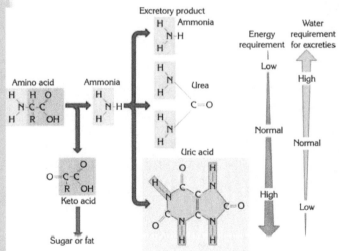

Insects, reptiles and birds all excrete this nitrogenous compound. Since this substance is insoluble in water and is excreted together with only a small amount of water, animals excreting uric acid are well-adapted to terrestrial life. Wastes from animals are excreted from different organs of the body. The excretory organs and the waste for which they are responsible for removal are as follows:

Kidneys : H_2O, urea, salts and other substances, such as medicines

Lungs : Carbon dioxide and water

Intestine : A small amount of water and indigestible material, for example cellulose

Skin : Water, a small amount of urea, and salts.

2. The Human Excretory System

The human excretory system is composed of kidneys, a urinary tract or ureter, urinary bladder, and urethra

a. The Kidney

The kidneys are two bean-shaped organs situated in the lower thoracic region of the back. They are between 120-150 g in weight and are protected by a thick layer of fat. The upper region of each kidney is covered by an adrenal gland.

Each kidney is supplied with blood from a branch of the renal artery. Filtered blood leaves through a branch of the renal vein. Urine formed in the kidneys drains into the urinary bladder before being expelled.

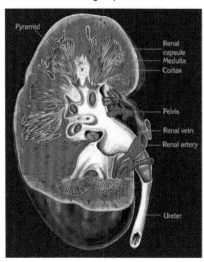

Figure: A cut-away section of a human kidney, showing the details of structures in both longitidunal and transverse section

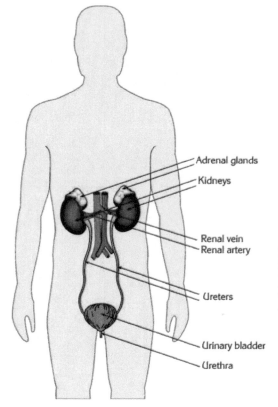

Figure: The structures that comprise the excretory system, from the filtration of urine to its expulsion from the body.

> ! Ammonia is the primary nitrogenous waste product, but it is highly toxic.
>
> It is the excretory substance of freshwater organisms.
>
> Humans produce urea as an excretory product.

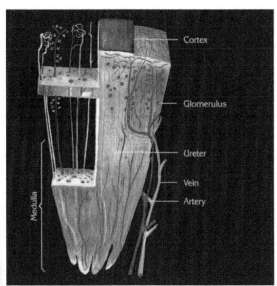

Figure: A three-dimensional segment through a kidney, showing individual nephrons.

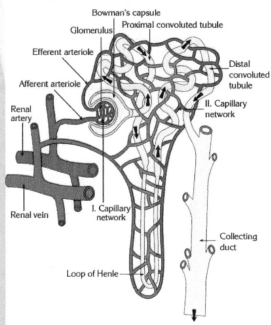

Figure: The structure of a nephron, detailing the path of blood flow and filtered substances.

It is composed of three main parts

Renal Cortex: It is red in color and contains the Malpighian bodies, comprising the Bowmann's capsule and glomerulus, which give it its rough structure.

Renal Medulla: It is located directly beneath the cortex. Urinary tracts which drain from the cortex form pyramids in this region.

There are approximately 8-10 laterally arranged Malpighian pyramids. The apex of each pyramid is located in the medulla and its base is located in the cortex.

Renal Pelvis: It forms the innermost portion of the kidney. Its function is the collection of urine from the Malpighian pyramids, the site of 15-20 orifices opening into the pelvis. The pelvis transmits the accumulated urine to the ureter.

2. The Structure and Function of the Nephron

Kidneys are composed of units known as nephrons. In each kidney there are approximately 1200 nephrons. Nephrons filter approximately 180 liters of fluid and form 1.5 liters of urine per day.

A nephron consists of three units:

Glomerulus   Bowmann's capsule   Urinary tract.

a. Bowmann's Capsule

It is a U-shaped, semi-spherical structure. The inner surface consists of squamous epithelial cells. The Bowmann's capsule forms the tip of the nephron.

b. Glomerulus

It is a ball of arterial capillaries located in the Bowmann's capsule. Each glomerulus is formed by capillaries from a branch of the afferent renal arteriole. These capillaries exit the Bowmann's capsule as a single structure, the efferent arteriole. It is the only example in the body of capillaries between two arterioles, in direct contrast to the other capillaries. The blood pressure is high in the capillaries of the glomerulus due to its position between two arterioles.

As a result, filtration is very effective. They also differ structurally from the other capillaries as their capillary wall is twice as thick as that of other capillaries.

This feature enables them to resist high blood pressure and to prevent loss of proteins and leucocytes. Only water and water-soluble molecules pass through the Bowmann's capsule from the glomerulus, and there is no reabsorption as in the other capillaries.

c. Malpighian body

It comprises a Bowmann's capsule and glomerulus.

Beneath the Malpighian body is the proximal convoluted tubule. It is formed from cuboidal epithelial cells. The proximal convoluted tubule extends into the loop of Henle and then into the distal convoluted tubule. The total length of these tubules that constitute a nephron in humans is approximately 5 cm.

d. Adrenal Veins

The renal artery supplying each kidney is a branch of the aorta. It forms 3-5 branches in both the left and the right kidney. These branches pass through the pyramids and merge between the medulla and cortex to form an arterial band.

The arteries flowing away from this band branch into afferent arterioles which enter each glomerulus to form highly branched capillaries. One arteriole leaves each glomerulus as an efferent arteriole, which rebranches into capillaries to form a network surrounding the tubule.

Blood flows from these vessels to the venules which merge to form a ventral band between the medulla and cortex. Venules extending from this band pass through the pyramids to the pelvis. They finally exit the kidney as the renal vein.

The renal artery transports blood rich in oxygen and metabolic wastes, whereas the renal vein transports blood that is rich in carbon dioxide with only a small amount of metabolic waste.

b. Urine Formation

There are three phases during urine formation;

filtration   reabsorption   secretion

1. Filtration

Urine formation begins with glomerular filtration of water, various ions, amino acids, sugar and the nitrogenous wastes. These substances pass to the Bowmann's capsule from the glomerulus. Blood cells, plasma proteins and lipid molecules however, remain in the glomerulus. Thus the fluid which passes into the Bowmann's capsule is proteinless plasma. Every day, 180 liters of fluid is filtered from the glomerulus into the Bowmann's capsule.

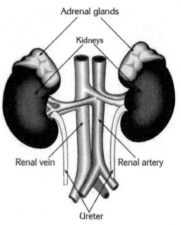

Figure: The position of the kidneys in relation to their blood supply.

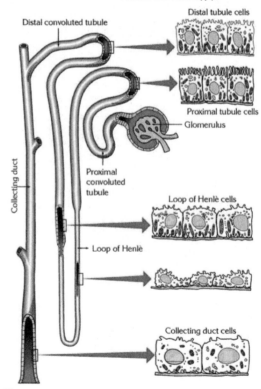

Figure: A nephron contains a variety of epithelial cells adapted for their specific function of absorption and filtration.

Blood is filtered between glomerulus and Bowmann's capsule.

Needed substances are reabsorbed through the tubules of the nephron. At the end, urine is formed and excreted into the urinary bladder via the ureters.

An equivalent volume of 1/4 of the blood pumped through the heart during one cycle passes through the kidney. Blood is filtered physically in the glomerulus by the action of blood pressure.

2. Reabsorption

It has been proven that all amino acids, glucose and some urine is reabsorbed at the proximal convoluted tubule. Sodium, chloride and bicarbonate ions are reabsorbed and are taken up by cells by active transport. However, 99% of water is reabsorbed passively from different regions of the tubule.

a. The proximal convoluted tubule

Reabsorption plays an important role in urine formation. Microvilli in the cells of the tubule walls increase the available surface area for absorption. A great number of mitochondria are present in the microvilli region, indicating a high level of metabolic activity. Thus, all materials, with the exception of water and urine, enter the cells. These reabsorbed materials are transferred back to the capillaries which surround the tubules. If an organism lost its ability to reabsorb essential filtered materials, death would occur rapidly.

b. Loop of Henle

It is specialized to produce a high concentration of sodium chloride in the medulla. The walls of the descending (first part) loop of Henle are relatively permable to water but impermable to sodium and urea. As the filtrate passes down the loop of Henle, water moves out by osmosis. This concentrates the filtrate in the loop of Henle. At the turn of the loop, the walls become more permable to salts and less permable to water. As the concentrated filtrate moves up in the loop, salt diffuses out into the interstitial fluid.

c. Distal convoluted tubule

It has an active role in moving molecules from the blood into the tubule, a step in urine formation called tubular secretion. Secretion of certain substances is controlled by an adrenal cortex hormone, aldosterone.

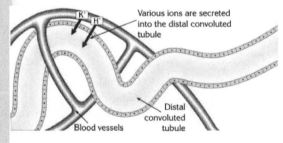

Various ions are secreted into the distal convoluted tubule

Distal convoluted tubule

Blood vessels

Figure: The distal convoluted tubule is the site of the secretion of penicillin, ammonia, potassium and pigment. Thus, blood pH is regulated.

d. Collecting ducts

They are located in the renal medulla. Antidiuretic hormone, or ADH, is secreted when the concentration of the blood increases. As a result, reabsorption of water from the collecting ducts back to the blood increases, thus concentrating the urine in the duct. Collecting ducts carry urine to the renal pelvis.

3. Secretion

The cells of the distal convoluted tubule excrete molecules such as ponicillin, ammonia, hydrogen, potassium, pigments and excess acids

The content of urine: Approximately 1-1. 5 liters of urine are produced per day. Its pH fluctuates between 5 and 7, and it contains the following substances in the following proportions:

3% organic molecules (urine, uric acid, creatine)

2% mineral salts (sodium, potassium, calcium, chloride and phosphate).

A small amount of leucocytes and epithelial cells. The remainder is water.

Urine is transported into the urinary bladder by the ureter of each kidney. The urinary bladder normally stores 300-500 cm^3 urine, and its wall can stretch to accommodate amounts of urine in excess of 300 cm^3. Sensory neurons in the innermost layer transmit impulses to the brain that a volume of urine has collected in the bladder. The central nervous system than stimulates the smooth muscle of the bladder wall to contract and urine is expelled from the body.

The urinary pathway

Each ureter resembles a pipe for urine expulsion originating from the urinary bladder. In men it passes through the prostate gland and extends out through the penis. In women it passes from the urinary bladder through the urethra and is expelled.

Kidneys → ureter → urinary bladder →urethra
formation   transport   storage   excretion

c. The Regulatory Function of the Kidneys

1. Regulation of filtration

The juxtaglomerular apparatus adjusts the pressure of filtration and the final composition of urine. This system is located in a specialized region of the distal convoluted tubule. The convoluted portion of the distal tubule is adjacent to the afferent arteriole.

The juxtaglomerular apparatus releases renin if the blood pressure in the afferent arteriole falls. Renin is an enzyme that changes angiotensin, a plasma protein made by the liver, to angiotensin I. Angiotensin I is changed to angiotensin II by a converting enzyme found in the lining of the pulmanory (lung) capillaries.

Angionsion II, a powerfull vasoconstrictor, stimulates the adrenal cortex to release aldosterone into the blood. The blood pressure thenrises as a result of vasocontriction and reabsorption of sodium ions (Na$^+$), followed by water.

Atrial natriuretic hormone (ANH) is secreted by the atria of the heart when cardiac muscle cells are streched due to increased blood volume. ANH inhibits the secretion of renin by the juxtaglomerular apparatus and the secretion of aldosterone by the adrenal cortex. As a result, more Na$^+$ is excreted. When Na$^+$ is excreted, so is water, and therefore, blood volume and blood pressure decrease

In human males, urethra passes through the penis, while in females, urination is performed directly from the urinary bladder via a short urethra.

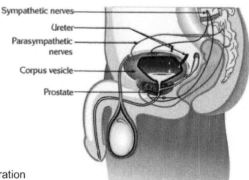

Figure: The structure of the urinary bladder and the structures that regulate its function in a male.

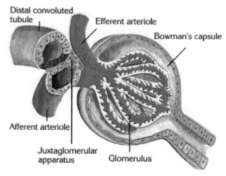

Figure: In the regulation of filtration, the juxtaglomerular apparatus secretes renin. The target structures of renin are the efferent arterioles which constrict, increasing the filtration rate at the glomerulus.

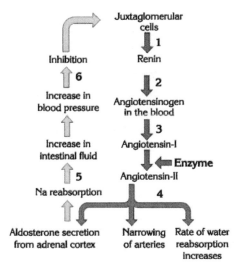

Figure: Factors influencing the reabsorption of water and sodium ions.

1. Juxtaglomerular cells secrete renin.
2. This triggers the accumulation of angiotensin I in the the blood.
3. Angiotensin I is converted to angiotensin II by the action of a converting enzyme.
4. Under the influence of angiotensin II, more water and sodium are absorbed by the kidneys. The arterioles are also narrowed
5. The secretion of aldosterone increases the rate of Na reabsorption and the volume of interstitial fluid. It also increases the blood pressure.
6. As the blood pressure rises, by a negative feedback system, the secretion of renin from juxtaglomerular cells ceases.

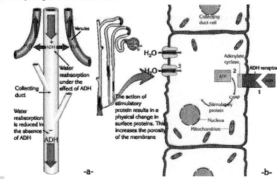

Figure:

a. When the body needs to conserve water, ADH production results in increased water reabsorption when the osmotic potential of the blood increases. ADH secretion is inhibited, resulting in dilute urine when the osmotic potential of the blood decreases.

b. The action of the peptide hormone ADH on the cells of collecting duct.

1. ADH binds to receptor proteins.
2. Cyclic AMP activates stimulatory proteins.
0. The opening of protein pores allows water molecules to move across into the collecting duct cells.

2. Regulation of water

The water level of the body is regulated by ADH, or antidiuretic hormone, also known as vasopressin, secreted from the posterior lobe of the pituitary. Its secretion, however, is regulated by the hypothalamus. ADH increases the reabsorption of water from the collecting duct. The excess water increases the amount of water in the blood, and subsequently the blood pressure increases. Consequently, glomerular pressure increases and more water is filtered. An increase in blood and tissue fluid, however, inhibits ADH secretion.

As a result, the permeability of the tubule decreases, and water reabsorption decreases. Homeostasis is restored. Factors affecting the quantity of water reabsorbed include the concentration of salts and sugars in the glomerulus. A high concentration results in less water reabsorption. This phenomenon accounts for the increase in the amount of urine excreted by diabetics.

A single kidney is capable of taking over the function of the other if one is lost from the body. Its size however, increases to compensate for the absence of the other. In reality only 25% of a normal functioning single kidney is necessary for filtration of urine from the blood in humans. If the kidneys fail to filter the blood sufficiently, a condition known as uremia results.

3. Salt : water balance

Salt reabsorption from the tubules is achieved by the action of aldosterone secreted from the adrenal glands. Since water molecules are held within the body by sodium (Na^+) ions, their insufficient reabsorption from the proximal convoluted tubule results in reabsorption of water from the distal convoluted tubule and collecting duct. The excess Na^+ in the body results in excess water accumulation and consequently edema.

Approximately 620 g of chloride (Cl^-) are filtered from the glomerulus. Of this quantity, 500 g is reabsorbed from the proximal convoluted tubule, 110 g is reabsorbed from the loop of Henlé, and 10 g is excreted within the urine. Potassium is reabsorbed from the proximal convoluted tubule, whereas some is released into the urine by the cells of the distal convoluted tubule.

Clearly, the major function of the kidneys is in the regulation of the osmotic potential of the blood and its material composition. In extreme cases however, the kidney cannot maintain the osmotic equilibrium of the blood. If, for example, an individual drinks a large quantity of sea water, his kidneys can only filter 2% of the 3% salt present. As his blood becomes more concentrated, water flows from the tissues to the blood. The victim consequently loses 0.5 liters of water for each liter of sea water drunk. The subsequent water loss from his tissues results in death. The body tissues of fresh water fish are hypertonic to their external environment. This problem of differing salt concentrations must be overcome in order for the organism to live. Fish use the following strategy. Water is taken up by osmosis. and the necessary salt ions are actively taken into the blood circulation by the cells of the gills. Since the body is covered with scales and skin which are impermeable to water, the fish can maintain its osmotic control. However, water taken up by osmosis from the gills

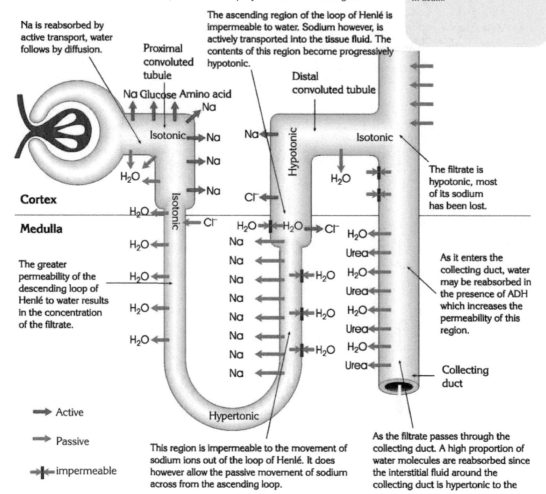

Na is reabsorbed by active transport, water follows by diffusion.

The ascending region of the loop of Henlé is impermeable to water. Sodium however, is actively transported into the tissue fluid. The contents of this region become progressively hypotonic.

Proximal convoluted tubule

Distal convoluted tubule

Na Glucose Amino acid

Cortex

Medulla

The greater permeability of the descending loop of Henlé to water results in the concentration of the filtrate.

The filtrate is hypotonic, most of its sodium has been lost.

As it enters the collecting duct, water may be reabsorbed in the presence of ADH which increases the permeability of this region.

Collecting duct

Active

Passive

impermeable

Hypertonic

This region is impermeable to the movement of sodium ions out of the loop of Henlé. It does however allow the passive movement of sodium across from the ascending loop.

As the filtrate passes through the collecting duct. A high proportion of water molecules are reabsorbed since the interstitial fluid around the collecting duct is hypertonic to the

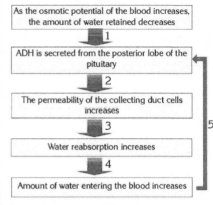

As the osmotic potential of the blood increases, the amount of water retained decreases

↓ 1

ADH is secreted from the posterior lobe of the pituitary

↓ 2

The permeability of the collecting duct cells increases

↓ 3

Water reabsorption increases

↓ 4

Amount of water entering the blood increases

5

Figure: The regulation of ADH secretion
1. The concentration of the blood increases.
2. This stimulates ADH secretion from the posterior lobe of the pituitary.
3. The target area of ADH is the cells of the collecting duct. The permeability of these cells increases.
4. Thus, water molecules move easily back into the venule.
5. As a result, the concentration of the blood decreases; blood volume increases.

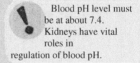

Blood pH level must be at about 7.4. Kidneys have vital roles in regulation of blood pH.

can not be prevented. To counteract the inward flow of water, Malpighian bodies actively pump the excess water out of the gills. Well-developed glomeruli assist excess water removal by filtration. Freshwater fish excrete high amounts of water due to insufficient reabsorption and have no need to drink water. In contrast, marine bony fish are hypotonic to their environment and their body fluid is less concentrated than sea water. For this reason, water loss from their tissues and excess salt uptake can not be prevented. To counteract the outward flow of water the following strategies are used:

The body is partially surrounded by a water-impermeable skin. Water is continuously drunk to prevent loss. Salts ingested with water are expelled by specialized cells in the gills by active transport. Nitrogenous wastes are excreted in the form of ammonia from the gills by active transport. The glomeruli of the kidney are small and tubular in form since there is little requirement for water to be excreted. No urine formation takes place in their kidneys, nor osmotic regulation of salt excretion.

4. pH Regulation

Kidneys have an important function in the regulation of the pH balance of the blood. Human cells, as is the case for most other animal cells, function only within a very narrow range of pH variation. For instance, the pH of human blood is approximately 7.4. Any fluctuation below 7 and above 7.7 results in death. The continuation of life processes is only possible between 7 and 7.7.

The kidneys balance the internal environment by secretion of acids or bases in response to any changes. They regulate pH by the manipulation of Na^+ and H^+ of H_2CO_3. Concentration of H^+ is equivalent to the concentration of HCO_3^- (bicarbonate) found in the distal tubule. If the extracellular fluid increases in acidity, the concentration of ions in the filtered solution decreases. Excess H^+ ions combine with buffers such as phosphate, and are expelled from the body.

If the extracellular fluid becomes alkaline, the concentration of HCO_3^- ions in the tubule exceeds the concentration of H^+ ions secreted from tubular cells. HCO_3^- ions are excreted in the form of $NaHCO_3^-$ (sodium bicarbonate), which combines with Na^+ ions. As a result, the pH balance returns to normal. The recycling of nitrogenous metabolic wastes

All living things excrete nitrogenous wastes. The amount excreted by one organism may appear negligible. However, when the volume of excretory products of a whole population is considered, it represents a large potential loss of nitrogen from the environment unless steps are taken to replace that which is lost. In a balanced, undisturbed ecosystem, nitrogenous wastes are returned to the soil where bacteria decompose them to soluble, usable forms. For large urban human populations, the same process is not followed. Instead, nitrogenous wastes are pumped into the sea and rivers, causing contamination and the proliferation of toxic algal blooms. The intensive farming techniques currently used aggravate the problem by the release of concentrated nitrogenous compounds into the water.

The important tumors of the kidney are the hypernephroid tumors (renal cell carcinoma) (Grawitz tumor), embryonal adenosarcoma (Wilm's tumor), and the papillary carcinomas which form in the parenchyma (adenopapilloma) or in the renal pelvis (transitional cell papilloma). The first two are relatively common, the last named are rare.

Hypernephroid tumors occur in adults, usually on the upper or lower poles of the kidney. They may be very large, but usually vary from the size of an olive to an orange. They usually appear orange in color when removed, and are surrounded by a fibrous tissue capsule. Microscopically the cells are clear and large with a cytoplasm containing liquid globules. They resemble the adrenal cortical cells in appearance. They are arranged in sheets and cords resting on a delicate connective tissue stroma, which encloses capillary strands. The tumors are malignant and metastasize to distant organs, usually by invading the renal vein and the vena cava. The bones and lungs may show the first clinical sign of involvement. Hematuria and pain in about one-third of the cases are other clinical features.

Adenoma and Adenocarcinoma. Small epithelial tumors of the kidney in the form of circumscribed tubular adenoma or intracystic papilloma may be found at necropsy. They rarely measure more than a few millimeters in diameter. Occasionally tumors of this type grow to large size and assume the morphology of papillary adenocarcinoma. These malignant growths have a much better prognosis than the more common hypernehproid carcinoma.

Wilms' tumor, or embryonal adenosarcoma, usually occurs in children from the ages of one to three. They are large, invasive tumors growing from either pole of the kidney or near the pelvis. They enlarge rapidly. Grossly, they are gray-yellow in color. Microscopically there is a highly variable appearance. The mixed or teratoid type contains skeletal muscle, glandular tissue, fat, bone, cartilage, etc. This type invades locally, involving the liver, pancreas, spleen, ureter, intestine and contiguous structures. The sarcomatous type is composed of malignant connective tissue, which encloses scattered embryonic renal tubules.

READ ME

Toxic wastes accumulate in the body if the kidneys fail. Consequently, cells and organs begin to deteriorate and eventually die. The toxic waste in the blood is removed by the successful use of artificial kidneys.

A tube is implanted into an artery and the blood is taken up to the pump. The waste-containing blood is pumped into the dialysis tubes, located within the dialysis fluid.

When the blood is passing through the dialysis fluid due to gradient difference, the purified blood returns to the circulation by a vein. The artificial kidney is like an organ of the body during this process.

The temperature of dialysis fluid is important for maintaining homeostasis.

The fresh dialysing solution is warmed up to the body temperature before giving it to the apparatus. The membrane of dialysis tubes is permeable to all materials except proteins and red blood cells.

Glucose may diffuse from dialysis fluid to the blood if its concentration is high. The body may be nitrified by adding nutrients to the solution.

A course of purification by artifial kidney takes about 6 hours. A patient suffering from kidney failure receives this course two or three times a week.

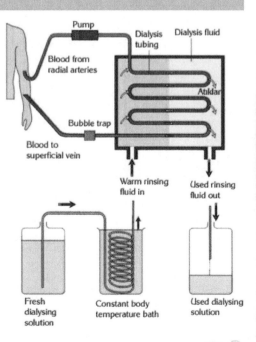

Excretory System

DISEASE AND DISORDERS OF THE EXCRETORY SYSTEM

Cystitis (Bladder infection)

Cystitis is an inflammation of the lining of the bladder. It is more common in women than in men since the passage of entry of bacteria up the urethra into the bladder is easier due to moister conditions.

The symptoms of cystitis include a frequent urge to urinate or pass water, pain or a burning sensation when passing water, darker urine that is sometimes blood stained, stomach ache and fever. The symptoms of cystitis can be relieved by taking the following steps. The sufferer should be advised to drink at least 3 liters of water a day to flush the bacteria out of the bladder and to avoid acidic drinks, such as fruit juices or alcohol. If the infection persists, a course of antibiotics usually clears the infection within a few days. Cystitis can best be avoided by keeping a high level of personal hygiene, drinking plenty of water every day and not overstraining the bladder.

Acute Renal Failure

Renal failure results usually from the failure of sufficient blood to reach the kidneys. Reasons for this include serious blood loss, cardiac failure or thrombosis. Blood poisoning due to arsenic or industrial toxic materials also results in total cessation of renal function. Unless the patient is hospitalized and placed on a dialysis machine, the condition may be fatal.

Phelonephritis

This common kidney disease is the result of the spread of *ESCHERICHIA COLI* bacteria to the urinary tract. It affects women more than men since contamination of the urethra is easier as the rectum and urethra are in close proximity to each other. As the bacteria spread up into the kidneys, a high fever develops. Other symptoms are back pain in the lumbar region and pain on passing urine, which is discolored.

A course of suitable antibiotics is prescribed to the patient.

Kidney Stones

Most kidney stones are formed from precipitated calcium salts. They can be deposited anywhere in the kidneys and produce intense pain. The reasons for formation of kidney stones are thought to be due to infection or obstruction, excessive ingestion of calcium or vitamin D, overexposure to sunlight and prolonged dehydration.

Patients with kidney stones are advised to drink large quantities of water and to decrease the level of calcium in their diet. For larger stones, shock-wave lithotripsy will shatter the stones to a size that can be passed out of the body in the urine.

Diabetic Capillary Sclerosis

Approximately 67 % of patients with diabetes mellitus die in coma or with infliction, or of vascular disease characterized by coronary occlusion or peripheral gangrene. In about 16 % of the cases, spherical or diffuse hyaline lesions are found in the glomeruli of the kidneys. These lesions are due to thickening, splitting and fusion of the inner capillary basement membranes. They do not originate from intercapillary tissue. It is seldom, however, that clinical symptoms are produced. When they occur, they are usually characterized by albuminuria, edema and hypertension. Bell, however, believes that in these cases most of the symptoms can be assigned to cardiac failure rather than to the renal lesions.

SELF CHECK

EXCRETORY SYSTEM

A. Key Terms

Excretion	Glomerulus
Kidney	Loop of Henle
Nephron	Afferent arteriole
Tubular secretion	ADH
Renin	Urinary bladder
Urethra	Urea

B. Review Questions

1. What is excretion?

2. What are the main parts of a kidney? Explain the function of each.

3. List the organs involved in excretion.

4. What are the differences between glomerular capillaries and others?

5. Compare the location and function of afferent and efferent arterioles of a nephron.

6. Compare the renal artery and renal veins.

7. List and discuss the main steps of excretion.

8. Explain the functions of hormones involved in the regulation of excretion.

C. True or False

1. The main function of the kidneys is filtration of metabolic wastes from the blood.

2. The renal medulla is the outermost layer of the kidney.

3. The first step in urine formation is filtration of blood, which is done between the glomerulus and Bowmann's capsule.

4. The collecting duct is the part of the nephron where a high rate of water reabsorption occurs.

5. Returning of substances back into the body from the nephron is called reabsorption.

D. Matching

a. Urea   ( ) Primary nitrogenous waste of humans derived from amino acid breakdown.

b. Urinary bladder   ( ) Part of the nephron lying between the proximal and distal convoluted tubules that functions in water and sodium reabsorption.

c. Distal convoluted tubules   ( ) Organ where urine is stored before being discharged.

d. Glomerulus ( ) Highly coiled part of the nephron that is far from Bowmann's capsule, where tubular secretion occurs.

e. Loop of Henle   ( ) A cluster of capillaries surrounded by Bowmann's capsule, where filtration occurs.

Ca^{++}, Na^{+} and K^{+} are needed for normal nervous system activity. Their concentration is regulated by kidney. Urination occurs under the control of the brain.

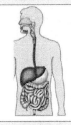
Digestive system provides nutrients to kidneys. While kidney remove wastes produced by organs of digestive system.

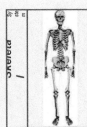
bone growth and repair. Kidneys help regulation of Ca^{++} which is needed for

Bones provide support and protection for internal organs including kidneys.

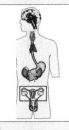
ADH secreted by pituitary and Aldosteron secreted by Adrenal cortex, help kidneys in reabsorbtion process. Kidneys keep blood concentration at a certain level so that transportation of hormones occurs normally.

Blood rids of metabolic wastes by means of kidneys.

Blood supplies nutrients to the kidneys.

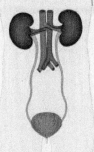

To maintain homeostasis, blood pH level must be sustained at a certain level. Lungs and kidneys together provide pH constant. Lungs excrete CO2 some of which produced by kidneys.

Muscular System

Kidneys have role in regulation of Na^{+}, K^{+}, and Ca^{++}, which are needed for muscular concentration.

Gall bladder is a muscular sac, which stores the urine.

GLOSSARY

A

acetylcholine: the most common neurotransmitter of the body. It is secreted at synapses.

action potential: the change in potential difference across the membrane of a neuron as the impulse passes during depolarisation.

active transport: the movement of a soluble substance across a membrane against a concentration gradient. This process requires energy.

adenine dinucleotide: a molecular unit of the DNA molecule comprising the purine adenine, a carbon sugar and a phosphate group.

adrenal cortex: the outer layer of an adrenal gland that produces cortisol, a hormone secreted in response to stress.

adrenal cortex hormone: a hormone produced by the anterior lobe of the pituitary. It stimulates the adrenal cortex to grow and secrete cortisol.

adrenal gland: a pyramid shaped gland that caps each kidney.

adrenal medulla: the central region of an adrenal gland that secretes epinephrine in response to stressful situations.

adrenaline: also known as epinephrine, it is a hormone secreted by the adrenal medulla. It stimulates the sympathetic nervous system.

aldosterone: a steroid hormone that regulates the excretion of sodium and potassium ions. It is secreted by the adrenal cortex.

alveolus: a one cell thick structure in the lungs through which gases are exchanged.

amoeboid cell: a single cell that is motile due to the movement of its cytoplasm.

amplification: an increase in intensity of for example, sound.

anemia: Either a deficiency of hemoglobin or of red blood cells.

antibody: an immunoglobulin. A specific protein molecule produced by plasma cells in order to deactivate antigenic molecules.

antidiuretic hormone: (ADH) see vasopressin.

antigen: a substance usually protein in nature that the body reacts to as an intruder.

artery: a blood vessel that generally transports oxygenated blood under pressure away from the heart. Its thick walls are adapted to the internal high pressure of the blood.

asphyxiation: the condition where insufficient oxygen is available to the lungs. A deficiency of oxygen in the blood can then lead to death if the level of oxygen is low enough. The brain is the most sensitive organ to oxygen lack.

atlas: the first cervical neck vertebra forming a site of attachment with the skull.

atria: the upper chambers of the heart.

autocrine gland: a gland whose cells secrete a hormone that act on the gland itself.

autonomic nervous system: the division of the nervous system that is responsible for maintaining homeostasis of the body.

axis: the second cervical neck vertebra located directly beneath the atlas vertebra.

axon: the component of a neuron that conducts an impulse away from the cell body. It is long and thin in structure.

B

beriberi: a disorder resulting from a lack of vitamin B. The right side of the heart becomes weakened and enlarged.

bile: a secretion produced by the liver that emulsifies fats during digestion.

bolus: a pellet of food formed by the mixing of masticated food with saliva by the tongue.

Bowman's capsule: a hollow sac of cells in the cortex of the kidney. It contains a glomerulus.

bronchus: a tube impregnated with rings of cartilage. It forms the connection between the trachea and a lung.

<hr>
C

calcitonin: a hormone secreted by the thyroid gland. It acts by lowering the calcium concentration of the blood.

canine: a sharp pointed tooth characteristic of carnivores, used for ripping and shredding meat.

capillary: a microscopic blood vessel. Its walls are only one cell thick. Tissue cells obtain oxygen and nutrients from capillaries.

cell body: the region of a nerve cell that contains the nucleus.

chemoreceptor: a sensory cell or organ that responds to a chemical stimulus. (see receptor)

cholesterol: a steroid forming part of animal cell membranes. Cholesterol in excess leads to formation of gall stones and thickening of the lining of arteries.

chondrocyte: a cell that constitutes cartilage tissue. It secretes the cartilage matrix.

cilia: short, hair-like protein structures that protrude from some cells. They are used for movement of materials across the cell surface or for locomotion.

cirrhosis: a serious disease of the liver affecting heavy drinkers. In this condition, the liver cells are destroyed and replaced by fibrous connective tissue. This disease can lead to liver failure.

<hr>
D

deamination: the first stage in the breakdown of amino acids. The amino group is removed and ammonia is formed.

decongestant: a drug which acts by reducing the amount of mucus present in the respiratory passageways.

dendrite: a component of a nerve cell. It is a long thin fibre which conducts an impulse towards the cell body. The membrane itself becomes less negative for a moment.

depolarisation: the sudden movement of potassium ions into a neuron through gates in the membrane.

diabetes mellitus: a disorder of carbohydrate metabolism where the glucose concentration of the blood is disrupted.

diaphragm: a muscular structure forming the floor of the thorax. It assists In Inspiration and expiration of air in and out of the lungs.

<hr>
E

effector: a muscle or a gland that produces a response to a nervous stimulus.

endocrine gland: a gland that secretes its products directly into the blood or into a tissue. It is a ductless gland.

epiglottis: a flap of tissue in the throat that prevents the entry of food into the trachea during swallowing.

exhalation: the expulsion of air from the lungs. This is achieved by the raising of atmospheric pressure by the upward movement of the diaphragm and the inward movement of the ribs.

exocrine gland: a gland that releases its secretions via a duct in the epithelium.

exophthalmic goitre: see external goitre.

external goitre: a contition where the thyroid gland responds to iodine deficiency by enlarging visibly.

<hr>
F

feces: insoluble waste products of digestion in animals.

fetus: the developmental stage of a vertebrate from the point where the major external features are formed to the end of pregnancy.

<hr>
G

gastric juice: a mixture of digestive substances produced by the stomach. It comprises hydrochloric acid, water and pepsinogen.

gastrin: a hormone secreted by the mucosa of the stomach in response to a quantity of food in the stomach. It stimulates the gastric glands to produce gastric juice.

glomerulus: a ball of capillaries located within each Bowman's capsule. Each is supplied by a branch of the renal artery.

goitre: see external goitre

grey matter: the region of the brain and spinal cord where cell bodies of axons are found.

growth hormone: a hormone produced by the pituitary gland under the influence of the hypothalamus. It stimulates body growth.

H

haversian canals: a component of bone tissue. They are microscopic channels located at the centre of each osteon. Capillaries and nerves run through these channels.

hemoglobin: a complex protein molecule present in the blood of vertebrates. It transports oxygen from the lungs to the tissues.

hemophilia: a disorder of the blood clotting mechanism. In this condition, a defect in one of the steps prevents clotting.

heparin: a chemical which functions within the blood as an anti-clotting agent. It prevents clotting from occuring in areas where it is not needed.

hepatic portal vein: a blood vessel of the body that supplies the liver with blood from the small intestine. Glucose and amino acids are transported through this vessel.

histamine: a chemical produced by most cells. It increases vasodilation and is a characteristic of allergic inflammatory reactions.

homeostasis: the maintenance of a steady state in the body by autonomic regulatory mechanisms.

hormone: an organic molecule produced in small amounts by specialised cells. It functions as a chemical messenger.

hypothalamus: a region of the brain that regulates the pituitary gland, the autonomic nervous system, water balance, body temperature as well as appetite and emotional responses.

I

incisor: a flat tipped tooth located at the front of the jaw. It is used for biting through hard food material.

inhalation: the intake of air into the lungs by the lowering of pressure, the downward movement of the diaphragm and the outward movement of the ribs.

interstitial fluid: a fluid that surrounds all cells. The cells obtain nutrients from this solution.

intrinsic factor: the substance present within the stomach responsible for the uptake of vitamin B_{12}.

Islets of Langerhans: the region of the pancreas responsible for the production of hormones. The alpha cells of the endocrine region secrete glucagon. The beta cells secrete insulin.

J

juxtaglomerular apparatus: specialized smooth muscle cells of the arteriole supplying the glomerulus of the kidney. It is the source of the enzyme renin which regulates blood pressure.

K

ketosis: an abnormal condition of the body where ketone bodies are produced in greater quantities than normal. This can result from diabetes mellitus, starvation or from a high-lipid diet.

L

lacunae: small hollows in cartilage tissue that form cavities. Each lacuna contains a chondrocyte or cartilage cell.

larynx: an organ adjacent to the upper region of the trachea. It contains the vocal cords.

leucocytes: ameboid white cells that defend the body from invading organisms.

lymph: a liquid that circulates around the body in lymph vessels. It consists of leucocytes which can develop into other types of cells such as macrophages when needed.

lysosome: a cell organelle containing enyzmes capable of hydrolysing the components of a cell. It is active during phagocytosis.

macrophage: a large cell formed from a monocyte. Its function is to phagocytose unwanted or foreign particles such as bacteria.

medulla oblangata: a structure of the brain controlling respiration blood pressure and heart beat.

melanin: a pigment produced by the cytoplasm of skin cells. It protects the skin aganist ultraviolet light.

meninges: the three membranes that surround the brain and the spinal cord.

negative feedback: the most common regulation of hormone production. As the concentration of a hormone increases, its production decreases.

nephridium: the organ of excretion of annelids such as an earthworm. Each segment of the body contains a nephridium.

neuron: a nerve cell which transmits an impulse from a receptor. It is usually composed of an axon, a cell body and dendrites.

Nissl substance: the structure within the cell body that is the site of protein synthesis.

obesity: the condition where an individual is seriously overweight.

offal: organs of the body of an animal that are considered edible. For example the kidneys, liver or brain.

osmosis: the movement of water across a membrane from a region where water is present in high concentration to a region where it is present in lower concentration.

ossification: the process of bone formation by the activity of osteoblasts.

osteocyte: a specialised cell formed from connective tissue. It is found within bone matrix and is a mature bone cell.

paracrine gland: a gland whose secretions affect a target tissue or organ in close proximity to the site of production.

parasympathetic nervous system: a division of the autonomic nervous system. It is active in returning the body to its normal state after stressful situations. Its function is to conserve energy.

parathormone: a hormone secreted by the parathyroid gland. It triggers the release of calcium ions from bones to the blood. It regulates the calcium concentration of the blood.

pathogen: any organism that that produces a disease in its host.

pepsin: an enzyme secreted by the stomach in gastric juice. It hydrolyzes proteins.

pericardium: a thin fibrous membrane surrounding the heart. Its function is protection of the heart.

perilymph: the liquid contained in the vestibular canal and the tympanic canal of the human ear.

periosteum: a membrane formed from connective tissue that surrounds the bone. It is capable of forming new bone itself.

peristalsis: the mechanism describing waves of contraction of smooth muscle in the esophagus. It is responsible for the movement of food to the stomach.

phagocytosis: the process where by a cell surrounds or engulfs another cell or food particle. The target structure is taken up by the cell and broken down.

pituitary: a small structure in the brain located directly beneath the hypothalamus. It is responsible for example, for the secretion of growth hormone and thyroid stimulating hormone.

progesterone: a hormone produced in females by the corpus luteum of the ovary. During pregnancy it is produced by the placenta.

prothrombin: a component of the clotting process. It is produced by the liver in the presence of vitamin K.

R

receptor: the component of a nervous system that receives a stimulus.

reflex: a nervous response to a stimulus. The response is automatic and usually not under the control of the individual.

resuscitation: an emergency procedure which involves

i) heart massage to stimulate the heart of a casulty to recommence beating.

ii) inflation of the lungs of the casulty who has stopped breathing using exhaled air from the first aider.

rickets: a disorder resulting from a lack of vitamin D. The bones of the body lose their rigidity and become bent (particularly those of the legs which support the body).

S

Shwann cell: a cell that wraps itself around a neuron forming an insulating myelin sheath.

scurvy: a disorder resulting from a lack of vitamin C in the body. The skin becomes weakened and bone growth is retarded.

sodium-potassium pump: the mechanism where by an impulse can be conducted along an axon.

somatotrophin: (see growth hormone)

sympathetic nervous system: a division of the autonomic nervous system. It is operative during periods of stress and prepares the body for emergency action by mobilizing sugar for energy.

synapse: the junction between either two neurons or between a neuron and an effector.

T

testosterone: a steroid hormone secreted by the testes.

U

urea: a highly soluble nitrogenous waste substance. It is produced by the liver of amphibia and mammals. It is excreted by organisms that do not need to conserve water.

uric acid: an almost insoluble nitrogenous waste substance produced from ammonia. It is formed by organisms that need to conserve water.

urine: the liquid waste of animals. It contains mostly water, urea, metabolic waste products and salts.

urogenital system: the term used to describe the combination of both the reproductive system and the urinary system.

V

vagus nerve: the 10th cranial nerve that provides communication between the brain and most internal organs. It forms part of the autonomic nervous system.

vasopressin: also known as antidiuretic hormone. It is secreted by the posterior lobe of the pituitary gland. It regulates the amount of water reabsorbed by the kidney.

vein: a blood vessel that generally carries deoxygenated blood under reduced pressure back to the heart. Its walls are thin due to the low pressure of the blood contents.

vena cava: a main blood vessel of the body into which deoxygenated blood drains back to the heart.

ventricles: the lower chambers of the heart.

villi: microscopic finger-like projections of the internal surface of the small intestine. Their function is the absorption of digested food material.

vocal cords: folded epithelium located within the larynx. When air passes over the vocal cords, they vibrate producing sound. They are a characteristic of most mammals.

Volkman canal: transverse channels running across the structure of compact bone. These channels allow blood vessels to penetrate the bone.

W

white matter: the portion of the brain and spinal cord composed mainly of axons.

INDEX

A

Acoustic axons, 48
Acidosis, 75, 184
ACTH, 65, 69, 76
Action potential, 20, 21
ADH, 70
Adrenal, 33, 75
  -cortex, 26, 27, 58, 75
  -medulla, 75, 77
Adrenal glands, 65, 69, 75, 77
Adrenaline, 33, 77, 118
Agglutination, 133, 140
Agranulocytes, 130
AIDS, 47,146
Albumin, 126
Aldosterone, 76
Allergens, 129, 144
Anastomoses, 121
Arteries, 11, 118, 156
Astigmatism, 44
Auditory nerve 45, 49, 50
Autocrine glands, 67
Axon, 17

B

Basophils, 129
Bicuspid, 113, 124, 147
Bile, 27, 176
Bilirubin, 129, 134
Bolus, 165, 167, 168, 173,
Bronchi, 33
Bronchioles, 151

C

Caecum, 130, 172
Calcitonin, 71, 73, 86
Calcium, 66, 71, 73
Capillaries, 11, 53, 112, 152, 167
Cardiac sphincter, 170
Cholera, 145
Cholesterol, 77, 122, 177
Chylomicron, 182
Chyme, 174, 175, 183
Clostridium tetani, 105
Coccyx, 28, 91
Colon, 34
Coronary arteries, 115, 118

Coronary veins, 115
Corpus callosum, 24, 35
Corpus ciliare, 39
Cortisol, 76

D

Diabetes mellitus, 73, 81
Diastole, 116, 119, 147
DPT, 145
Duodenum, 170, 171, 176, 177
Dura mater, 22, 27
Dwarfism, 69, 70, 81

E

E. coli, 186
Eosinophils, 129, 132
Endocrine glands, 13, 14, 16, 25
Enterokinase, 181
Epilepsy, 33, 34
Epiphysis, 67, 79
Erythrocytes, 126, 127, 128
Exocrine glands, 64, 67, 73

F

Facial bones, 90, 95
Fibrinogen, 126, 132, 146
Fibula, 87, 94, 95
Fovea, 39, 41
Frontal bone, 41, 90
Frontal lobe, 24, 59
FSH, 65,69

G

Gastric juice, 171, 173, 176
Gastrite, 176
Gigantism, 69, 86
Glycogen, 73, 75

H

Haversian canals, 85
Heme group, 127, 129
Hemoglobin, 106, 127, 128
Hepatitis, 145, 189
Homeostasis, 8, 9, 10
Humerus, 87, 94, 95
Hypermetropia, 43
Hypertension, 122
Hypotension, 122

I

Ileum, 171

Ilium, 93, 171
Influenza, 34, 160
Inner ear, 45, 46, 47
Insulin, 73, 75
Interferon, 137, 138
Iodine (I), 60, 126
Iris, 39, 40, 62

K

Kuppfer cells, 136, 177, 178

L

Lactose, 180, 193
Leucocytes, 89, 122, 126
LH, 65, 69, 77
Lipase, 179, 193
Lymph nodes, 125, 136
Lymph vessels, 124
Lymphocytes, 130, 136

M

Maxilla, 90
Meissner corpuscles, 54, 56
Meninges, 35
Metacarpals, 94
Middle ear, 46
Monocytes, 130
Myocardium, 102, 114
Myopia, 43
Myxodema, 81

N

Nasal bone, 90
Neutrophils, 129, 132

O

Occipital bone, 90
Orbit, 41
Outer ear, 45
Oxyhemoglobin, 156, 160

P

Pancreatic amylase, 180
Parathormone, 80, 86
Pelvic girdle, 93
Peristalsis, 169, 185
Phalanges, 94
Placenta, 80
Pleura, 161

R

Radius, 94

Residual air, 161
Retina, 41

S

Serosa, 170, 171, 172, 173
Serum, 141, 145, 146
Sinoatrial Node, 117
Sternum, 85, 87, 90, 93
Submucosa, 172, 185
Superior vena cava, 117
Sweat gland, 54
Systole, 116, 119

T

Temporal bones, 90
Temporal lobe, 24, 47, 59
Tendons, 98, 102
Thorax, 152, 153
Threshold level, 19,38
Thyroxine, 118
Tibia, 87, 94, 95
Tricuspid, 113, 147
TSH, 67, 69, 71
Tuberculosis, 107

U

Ulcer, 107, 130, 145
Ulna, 87, 94, 95
Uremia, 184,186

V

Vagus nerve, 118, 176, 179
Varix, 120
Vasopressin, 70, 80
Veins, 112, 114, 115
Vital capacity, 154, 161

W

Withdrawal reflex, 30

REFERENCE

1. Solomon E. P.; Berg R. L.; Martin D. W. Biology fifth edition. USA: Sounders college publishing, 1999.

2. Gardner/Simmons/Snustad. Principles of genetics. Eighth edition. John Wiley publication. 1991

3. Mauseth. Botany. Second editions. Sounders college publishing, 1995.

4. Shier D.; Butler J.; Lewis R. Sixth edition. Essentials of Human Anatomy and Physiology. WBC. McGraw-Hill. 1998.

5. Lewis R. Life third edition. WBC. McGraw-Hill. 1998

6. Guyton, A. Fizyoloji, cilt 1-2-3. Ankara: Guven Kitabevi Yayinlari, 1984

7. Arpaci O. The introduction to Biology. Istanbul: Zambak publication, 2003

8. Kenci B. Dogan M. Arpaci O. Zoology. Istanbul: Zambak publication 2004

9. Arpaci O. Ozet M. Heather J. E. Biology 2. Istanbul: Zambak publication 2000

10. Arpaci O. Ozet M. Heather J. E. Biology 3. Istanbul: Zambak publication 2000

11. Arpaci O. Ecology. Istanbul: Zambak publication 2003

12. Demirsoy A. Yasamin Temel Kurarlari, cilt-1/Kisim 1-2. Ankara: Metaksan 1985

13. Schraer. Stoltze. Biology, The study of life. Revised third edition. Allyn and Bacon publications.

14. Campbell. Reece. Biology, sixth edition. USA, Benjamin Cummings publications 2002.

15. Biology. Silver Burdett publications 1986.

16. Lawire, R., Chemistry for you, Stanley Thornes publishing. England 1996.

17. Texas science grade 7. New York, USA. McGraw-Hill. 2002

18. Science Explorer. Human Biology; Needham USA, Prentice Hall 2000

19. Superhuman BBC, 2003 England.

20. The human body 1-2 BBC, 2002 England.

21. Human senses BBC, 2003 England.

Printed by Amazon Italia Logistica S.r.l.
Torrazza Piemonte (TO), Italy

26158390R00128